Alaeddine Hmidi
Malek Jihene

Deep learning for image classification

Alaeddine Hmidi
Malek Jihene

Deep learning for image classification

Preliminaries and generalities

ScienciaScripts

Imprint

Any brand names and product names mentioned in this book are subject to trademark, brand or patent protection and are trademarks or registered trademarks of their respective holders. The use of brand names, product names, common names, trade names, product descriptions etc. even without a particular marking in this work is in no way to be construed to mean that such names may be regarded as unrestricted in respect of trademark and brand protection legislation and could thus be used by anyone.

Cover image: www.ingimage.com

This book is a translation from the original published under ISBN 978-620-3-42393-8.

Publisher:
Sciencia Scripts
is a trademark of
Dodo Books Indian Ocean Ltd., member of the OmniScriptum S.R.L Publishing group
str. A.Russo 15, of. 61, Chisinau-2068, Republic of Moldova Europe
Printed at: see last page
ISBN: 978-620-4-06141-2

Deep learning for image classification.

Preliminaries and generalities

Hmidi Alaeddine, Malek Jihene.

This book provides engineering students, research masters students, doctoral students, researchers and teacher-researchers, specializing in electronics or computer science, with the knowledge necessary to understand the basic concepts: Deep learning for image classification.

Table of Contents

Introduction

This book provides engineering students and Master students, specializing in electronics or computer science, with the necessary knowledge to understand the basic concepts: Deep learning for image classification.

In this book, we will provide the general basics of deep learning and classification, prerequisites that have been carefully chosen to allow a quick and easy understanding of the elementary concepts and the necessary contexts for deep learning and classification that will make this book relatively self-contained. In the first chapter, we will deal with neural networks. Then, in the second chapter we will present Deep Neural Networks. In what follows, we discuss and study interesting CNNs models for image classification and the optimization and compression methods and techniques used in FPGA-based deep learning acceleration. This work will end with a study of different implementation approaches already accomplished by determining their advantages and disadvantages. The last chapter represents recent trends in FPGA-based accelerators of deep learning models and could guide and direct future developments of efficient hardware accelerators and be useful for deep learning researchers.

Neural networks

Over time and since its early years, artificial neural networks have gone through alternating periods of excitement and oblivion, during which they have been known by various names and many neural models have been proposed. In this chapter, we review the most important successes of artificial neural networks. We will start from the Perceptron model proposed by Rosenblatt [1] in 1958, we will go through the Perceptron multilayers Until the proposal of a solution to the leakage gradient problem in the training of a deep neural network in 2006 by Hinton.

I. Formal neuron :

In 1943, Warren McCulloch and Walter Pitts presented formal neurons [2] considered as an abstraction of physiological neurons. They showed that simple formal neural networks can perform logical, arithmetic functions. This is a binary neuron, whose output is 0 or 1. In this model, the first operation performed is to calculate the sum of the magnitudes received as inputs, weighted by the synaptic coefficients, and a threshold is added to this magnitude. We note that each input is associated with a synaptic weight. Figure 1.1 represents the formal neuron architecture.

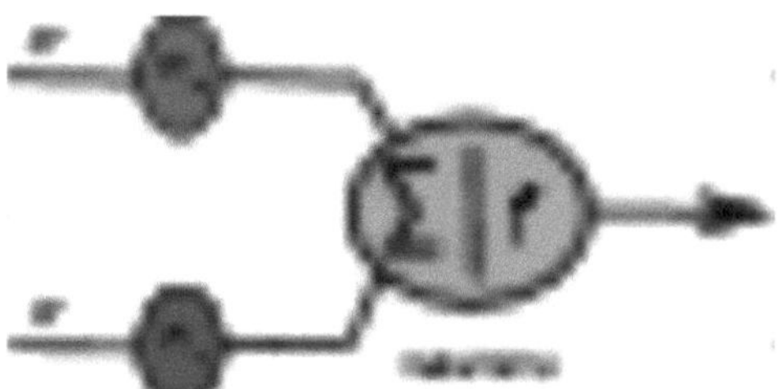

Figure 1.1 - Formal neuron architecture

II. Perceptron :

At Cornell University's Aeronautics Laboratory, the Perceptron was invented in 1958 by Frank Rosenblatt [1]. A Perceptron is a forward propagating neural network and a supervised learning algorithm of binary classifiers, separating two classes. The perceptron is very effective for linear classification functions. But difficulties arise if the two classes are linearly inseparable. To solve this problem, it is better to add more layers of neurons. Thus, the multilayer perceptron (MLP) appears to be a useful solution.

In a perceptron with N inputs and a single output s, each input $\vec{X}$ is associated with a synaptic weight $\vec{\omega}$. The operation performed consists in calculating the sum of the magnitudes received at the inputs, weighted by the synaptic weights, and to this magnitude is added a threshold (Heaviside function) f. The activation functions should preferably be strictly increasing and bounded.

$$p = \sum_{i=1}^{n} \omega_i \times X_i$$

The output s then results from threshold f

$$s = f(p) = \begin{cases} 1 & \text{si } p > 0 \\ 0 & \text{si non} \end{cases}$$

Figure 1.2 shows the perceptron.

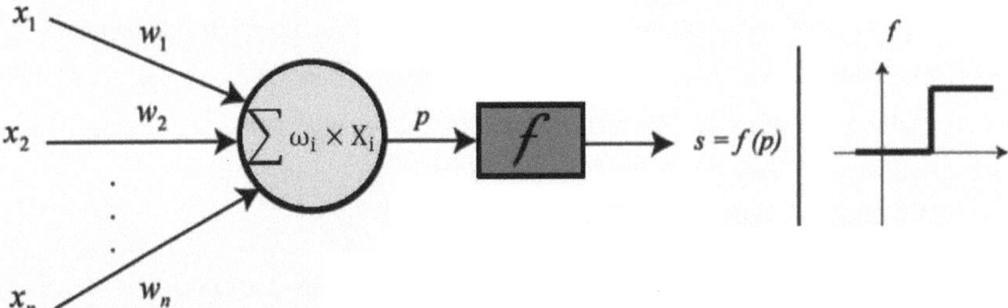

Figure 1.2 - Architecture of the simple perceptron.

III.   The multi-layer perceptron :

The multilayer perceptron (MLP) is a stack of single perceptron layers, or a sequence of single perceptron layers. The MLP is a forward propagating neural network, in which the incoming data flows from the input layer to the output layer. The PMC consists of N layers divided into three types of layers: the input layer which receives the input data to be processed. The output layer which provides prediction and classification and (N-2) hidden layers. The hidden layers are all the layers except the input and output layers which represent the real computational engine of the PMC. The PMC neurons are trained with the backpropagation learning algorithm. PMCs are designed to solve problems that are not linearly separable. Figure 1.3 shows the architecture of the multilayer perceptron.

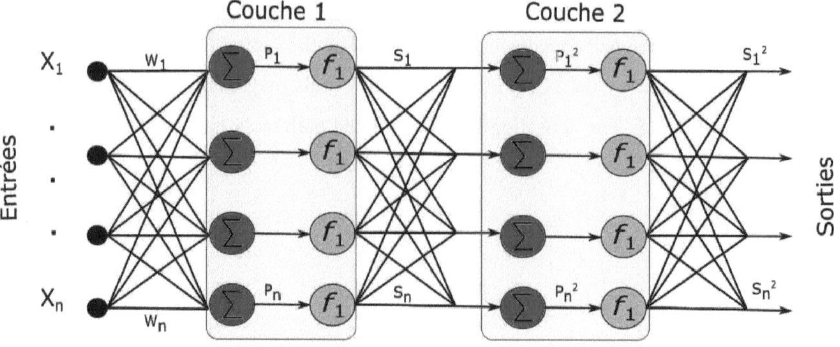

Figure 1.3 - Architecture of the multilayer perceptron.

We note that for k hidden layers :

$$s_n^k = f_2(p_n^k)$$

PMC networks are capable of computing a wider range of Boolean functions than networks with a single layer of computational units. However, the computational effort required to find the correct combination of weights increases significantly when more parameters and more complicated topologies are considered. The backpropagation algorithm was the most popular learning method that is capable of handling large learning problems.

IV.   Learning with backpropagation:

Gradient descent is an optimization algorithm that is essentially used to update the parameters of the training model. Updating this many parameters requires an operation that consists of optimizing the objective function that represents the total prediction error of the system on average for the whole training. The method that combines training and optimization is called gradient backpropagation. The back-propagation algorithm was originally introduced in the 1970s, but its importance was not fully appreciated until after a famous 1986 paper by David Rumelhart, Geoffrey Hinton, and Ronald Williams in a paper entitled "Learning representations by back-propagating errors" [3]. Back-propagation is the mechanism by which artificial neural networks learn. It is the most widely used learning algorithm in neural networks today. Backpropagation learning has become the most popular method of training neural networks. The reason for the popularity is the simplicity and its ability to be used for training non-linear networks of arbitrary connectivity. In addition, it works much faster than previous learning approaches, allowing artificial neural networks to be used to solve problems that were previously intractable. In simpler terms, after each forward pass through a network, the backpropagation algorithm performs a backward pass while adjusting the model parameters (weights and bias).

Backpropagation learning [3] has become the most popular method of training neural networks. The reason for the popularity is the simplicity and its ability to be used for training nonlinear networks of arbitrary connectivity.

The objective of the backpropagation algorithm is to define a parameter vector w that minimizes the error between the true labels (database) and the prediction labels in order to have a match between the true labels and the prediction labels. The calculation of the gradient of the error in this type of classification (the error between the true labels and the prediction labels) can be provided by the cost function. In general, the cost function can be written as an average over the training set. The goal of gradient runs is to iteratively converge to an optimized configuration of synaptic weights. The goal is to find a minimal configuration for the loss function by applying this formula:

$$\omega(t+1) \;=\; \omega(t) \;-\; \rho \cdot \frac{\partial C}{\partial \omega}$$

At the heart of the backpropagation algorithm is an expression for the partial derivative of the cost function C with respect to any weight ω (or bias b) in the network. It is often denoted by $:\frac{\partial C}{\partial \omega}$. The learning rate ρ determines the size of the steps we take to reach a local minimum. We note that a learning iteration refers to the passage of all the labels of the training base once in the network.

V. Convergence of learning :

When we develop a learning model, we try to teach it to achieve the desired goal: It may happen that the model finds difficulties and phenomena that hinder the learning process. This phenomenon is called "overlearning" or "overfitting". Overlearning is due to the formation of an overly complex model with a very large number of layers. In addition, overlearning also results from an insufficiently large training data set compared to the complexity of the training model. Regularization is a technique used to avoid it. A regularization is defined as a mechanism allowing to minimize the cost function by adding a regularization term to this cost function. It consists in penalizing the weights of too important connections. The importance of regularization is in the following two functions: minimization of overlearning and selection of a system whose weights will be the lowest in the case where many similar Ws can ensure the same classification of a set of data (same loss computed by these different Ws), all offering results that are most equivalent to the training data. The Dropout layer is another regularization solution, this layer will be described in the next chapter.

VI.   Alternatives to gradient descent :

1. Gradient descent:

Gradient descent [4] is one of the most popular algorithms to perform optimization and the most common way to optimize neural networks.

Gradient descent is a way to minimize an objective function (also called the cost or loss function) by updating the parameters in the opposite direction of the gradient of the objective function. Basically, there are three variants of gradient descent (GD) proposed in the literature: Batch gradient descent, Stochastic gradient descent, and Static gradient descent. These are Batch Gradient Descent (BGD), Stochastic Gradient Descent (SGD) and Minibatch Gradient Descent (MGD).

Each has its advantages and disadvantages in terms of accuracy and performance. For example, BGD gives the most accurate results, but requires a lot of extensive and expensive analysis on the data set. Also, contrary to the current understanding (that SGD is always the fastest), there is no single algorithm that outperforms the others at runtime.

2. Batch gradient descent :

Batch gradient descent is a variant of gradient descent that processes all training examples for each gradient descent iteration. For a large number of training examples, batch gradient descent can be very slow and computationally expensive. For this reason, it is preferred to use the gradient descent variants stochastic gradient descent or mini-batch gradient descent.

Batch gradient descent, calculates the gradient of the cost function for the training data set: The learning rate determining the size of an update performed. The batch gradient descent is guaranteed to converge to the global minimum for convex error surfaces and to a local minimum for non-convex surfaces.

3. Stochastic gradient descent (SGD)

Stochastic gradient descent, on the other hand, performs a parameter update for each training example. Stochastic gradient descent eliminates the redundant computations accomplished by batch gradient descent for large data sets by performing one update at a time. This redundancy achieved by batch gradient descent is due to the recomputation of gradients for similar examples before each parameter update. Stochastic gradient descent is much faster and can be used for online learning.

4. Gradient descent by mini-lots :

Mini-batch gradient descent performs an update for each mini-batch of N training examples. It reduces the variance of parameter updates, which can lead to more stable convergence. Common mini-batch sizes range from 50 to 256. They can be varied for different applications. Minibatch gradient descent is generally the algorithm of choice when training a neural network.

5. Gradient descent optimization algorithms:

Optimization is one of the key components that helps a model to train better during backpropagation when the weights are adjusted to minimize the loss error. The most popular optimization algorithms: SGD [5], RMSprop, AdaGrad [6], AdaDelta [8], Adam [7] . Momentum [5] is a method that helps speed up stochastic gradient descent (SGD). It is an approach that almost always enjoys better convergence rates on deep networks. Momentum adds a fraction of the previous update vector to the current update vector. RMSprop is an adaptive learning rate method first proposed by Geoff Hinton in Course 6 of the online course "Neural Networks for Machine Learning". It maintains a moving (updated) average of the gradients and uses this average to estimate the variance. Most deep learning frameworks include its out-of-the-box implementation.

RMSprop and Adadelta were both developed independently around the same time due to the need to solve Adagrad's drastically decreasing learning rates.

Deep neural networks

I. The value of deep architectures.

Multi-layer perceptrons (MLPs) were a widely used machine learning solution at the end of the last century. However, they are often used as classifiers and they are limited in their ability to process natural raw data. Features that are hand-crafted are needed. The development of an algorithm that uses predefined features that can be presented in an image like Eigenfaces [10], Kalman filter [11] or Viola-Jones [12]. The most difficult part of these algorithms is to define the key features and use them efficiently.

Unlike these algorithms (the feature definition), neural network algorithms must incorporate a stack of neural layers to automate the feature extraction process. Neural networks with more than three layers between input and output are called deep neural networks. Deep neural networks are based on artificial neural networks consisting of several layers of interconnected neurons that will be trained to be able to classify the incoming data. For image classification, convolutional neural networks can automatically extract features and topological properties of the image since they can act directly on the raw inputs (R, G, B) as opposed to different techniques and methods that benefit from different models based mainly on the construction of feature vectors from the raw inputs.   There are many common feature extraction techniques, among them HOG, SURF, LBP, and several types of neural architectures commonly used in various deep learning applications.

II. Convolutional neural networks (CNN)

Convolutional neural networks have always been the most powerful of all types of neural networks. They are Convolutional neural networks are biologically inspired networks that are widely used in computer vision for image recognition, object detection/localization and even text processing. They were first invented by Yann LeCun and others in 1998. The idea implemented by LeCun et al was older and based on work proposed in 1968 by David H. Hubel and Torsten Weisel in 1968, which earned them the Nobel Prize in Physiology and Medicine in 1981. This section is devoted to the convolutional neural network and its terminology. It will provide a brief overview of this architecture, which will be discussed in more detail later. Furthermore, it will not cover other types of deep learning.

1. Overview and basic structure of a convolutional network :

Convolutional neural networks (CNNs or ConvNet) are designed for input with a 2D structure in the form of an image or speech signal, and have always been the most efficient of all types of neural networks. They are defined for the presence of at least one or more convolution layers, pooling layers, fully connected layers as in a standard multilayer neural network and classification layers (Softmax). The basic structure of a typical CNN is shown in Figure 2.1 which incorporates the main operations in a ConvNet which are described as basic elements.

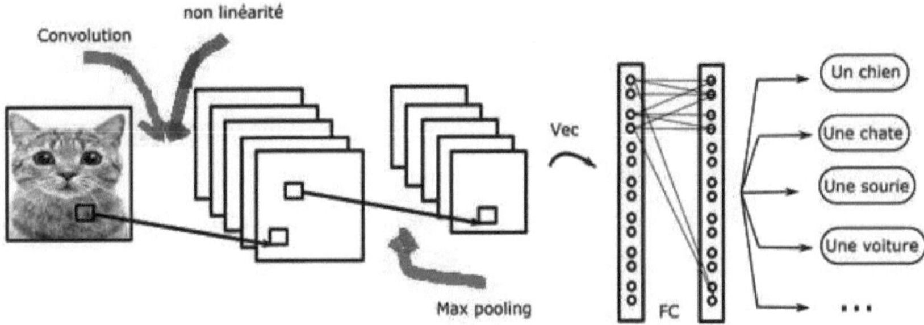

FIGURE 2.1 - Standard Convnet

A. Convolution layers and feature maps :

For convolution layers, a convolution operation is defined, in which a filter is used to map activations from one layer to the next. A convolution operation uses a three-dimensional weight filter with the same depth as the current layer. Convolution represents the main step and core of convolutional neural networks. A convolution is defined as the dot product between the parameters $F \times F \times D$ in the filter and the corresponding grid in the input volume. Usually F is an odd value.

The principle of convolutional layers consists in convolving images from previous layers in order to generate new output feature maps and extract features from the input image. The bias value is added to the scalar product. Note that each filter in a layer is associated with its own bias.

The convolution operation places the filter $F \times F$ at each position according to the number of stride[1] and pad[2] . Each position defines a spatial "pixel" in the next layer. The size of the next feature map is defined by the following formula:

$$Nouveau\ taille = \frac{((Taille\ de\ l'image - F + Stride + (2 \times Padding)))}{Stride}$$

Reducing the size of the image or feature map after the convolution operation is not desirable in general, as it tends to lose some information along the edges. This problem can be solved by using padding.

When we pass the filter along the input, we have to decide how many pixels we want to pass each time (S). (Stride: S) have the effect of rapidly increasing the reception field of each feature map in the next layer, while reducing the spatial footprint of the entire layer. It is most common to use a stride of 1, although a stride of 2 is also sometimes used. It is rare to use stride sizes greater than 2. Larger stride sizes can be useful in memory-limited environments or to reduce overfitting if the spatial resolution is unnecessarily high.

The number of filters (K) used in each layer controls the capacity of the model as it directly controls the number of parameters. In addition, increasing the number of filters in a particular layer increases the depth (number of feature maps) of the next layer. Each unique filter in a layer is associated with its own bias. The bias value is added to the scalar product. Using the bias simply increases the number of parameters in each filter by 1. The bias is learned during backpropagation. Each filter attempts to identify a particular type of spatial pattern in the image, and so the use of a large number of filters is necessary to capture a wide variety of possible shapes that are combined to create the final image. Assuming we have, for example, an incoming image of size 7x7 with pixel values of only 0 and 1 (this is a special case) and also a filter of size 3x3, the convolution can be calculated as follows in Figure 2.2.

---

[1] When we pass the filter along the input, we have to decide how many pixels each time we want to do it (S).
[2] Zero-padding (P): Sometimes it is convenient to fill the size of the input with zeros around the borders.

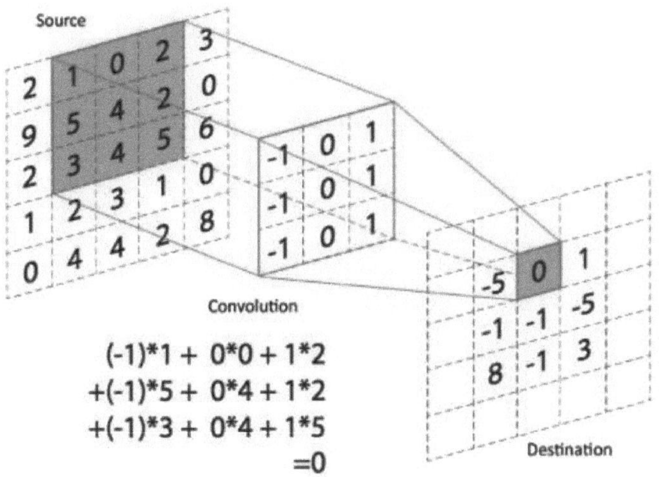

Figure 2.2: Example of convolution operations. The sliding window is 3 × 3 in size and moves 1 by 1 along the y-axis and x-axis.

For a 3D convolution, we can say that a convolutional layer accepts a volume of size :

$$W_1 \times H_1 \times D_1$$

and delivers a volume of size :

$$W_2 \times H_2 \times D_2$$

where:

$$W_2 = \frac{(W_1 - F + 2P)}{S}/S + 1.$$

$$H_2 = \frac{(H_1 - F + 2P)}{S} + 1.$$

$$D_2 = K.$$

To better visualize how a 3D convolution operation works, we report the convolution example presented in figure 2.3 :

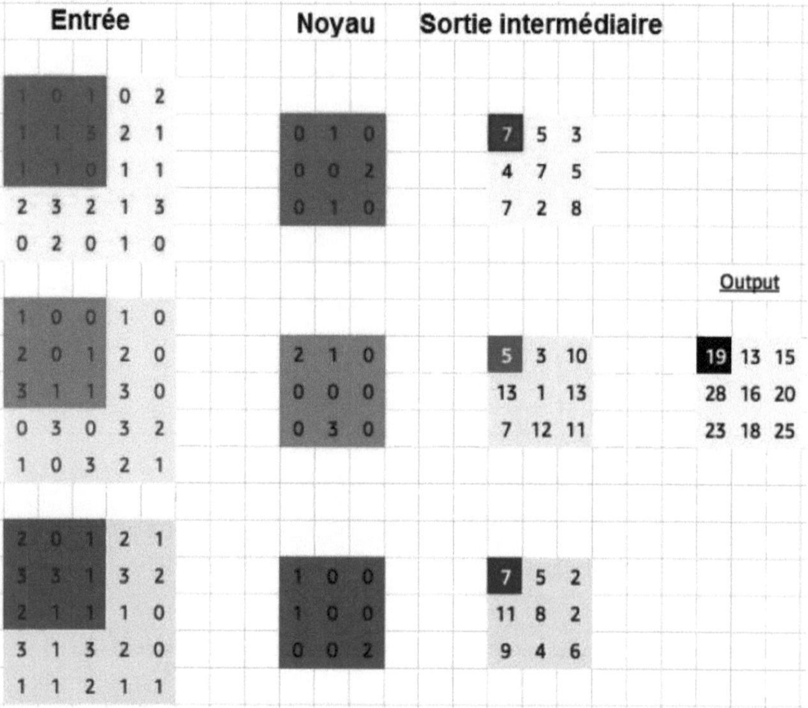

FIGURE 2.3 - Example of convolution operations, (W1 = H1 = 5), D1 = 3.

In this example, the input volume is of size equivalent to 5×5 (W1 = H1 = 5), D1 = 3 and P = 0. Here, we have a filter of size 3×3, each kernel is a 2D window of dimension 3 x 3 pixels and there are 3 channels in each kernel corresponding to each color channel (R, G, B). When the kernel is at a particular location on the 5×5 size input image with S=1 and P=0, its 3 components are multiplied with the corresponding channel pixel data to produce a single scalar number for each channel. So you get 3 scalars, one for each channel. Then these scalars are summed and another scalar representing the filter bias is added to the sum. The final result is a single scalar. The size of the output volume (in green) can be calculated by applying the formula seen previously:

$$W_2 = H_2 = \frac{(W_1 - F + 2P)}{S} + 1 = \frac{(5 - 3 + 0)}{1} + 1 = 3.$$

Reducing the size of the image or feature map after the convolution operation is generally not desirable, as it tends to lose some information along the edges. The convolution operation is followed by a nonlinear activation function layer. In the next subsection we will describe this layer

There are different types of convolutions as shown in figure 2.4.

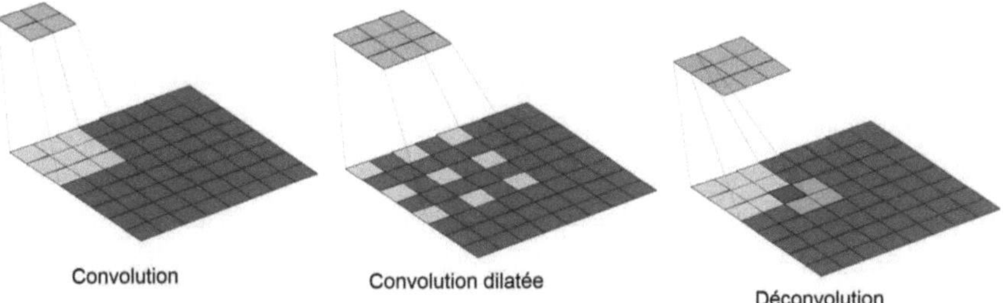

Figure 2.4 -The different types of convolution.

- **Classical or normal convolution/**

This is the widely known and used normal convolution, which we discussed in the first part of the paragraph.

- **Dilated convolution :**

This type of convolution is based on an additional hyperparameter called the expansion rate "d" in the appropriate layer. This hyperparameter indicates how much the kernel can be expanded. It is simply a matter of going through the image used for the convolution by skipping some of its pixels. The missing pixels in the kernel are replaced by "0". Expanded CNNs present interesting performances in different tasks such as synthesis and speech recognition.

- **Transposed convolution or deconvolution :**

The transposition of convolution in the context of CNNs is synonymous with deconvolution. The two types of convolution, presented in this section, are often confused. These two types of convolution are not very popular in the field of machine learning. Transpose convolutions are exploited in generative adversarial networks (GANs) and auto-encoders, or generally in any network that needs to reconstruct an image. The transposed convolutional layer, unlike the convolutional layer, is one of several strategies for performing oversampling. These types of convolution work with filters, kernels, strides like convolution layers but with an output larger than the input.

B. Non-linear activations:

The choice of the activation function is very important during the design of a CNN. Immediately after a convolution layer, we find a nonlinear unit [13,14] whose purpose is to introduce nonlinearity accurately, because most of the real-world data we would like our ConvNet to learn is nonlinear, instead, the convolution is a linear operation. The most popular nonlinear function is the rectified linear unit (ReLU), but it is not the only one. In the past, other functions were used such as the hyperbolic tangent (tanh) and sigmoid, but these functions are no longer exploited because they require complex arithmetic and present difficulties in hardware implementation. In addition, the hyperbolic tangent does not represent a solution for the gradient vanishing problem. The researchers found that the ReLU layers work much better because the network can be trained faster (due to computational efficiency) without a significant difference in accuracy. The function applies $f(x) = max(0, pixel)$ to all values of the input volume. Basically, this layer simply replaces all negative pixel values in the feature map with 0. An activation function does not change the dimensions of the layer because it is a simple one-to-one mapping of the activation values. Figure 2.5 shows examples of widely known activation functions:

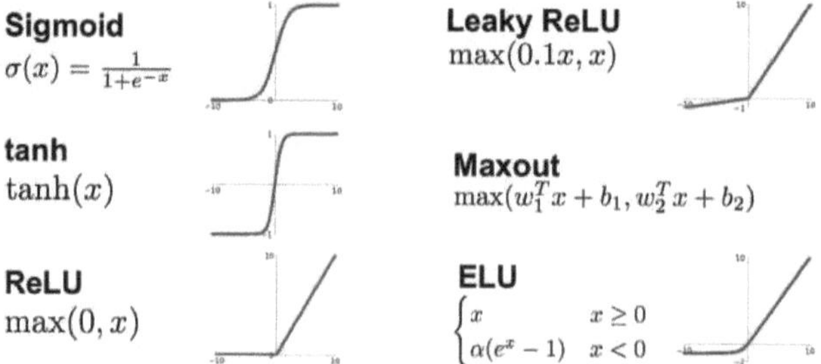

FIGURE 2.5 - Non-linear activations.

C. Sub-sampling / Pooling / Pooling :

It is common practice to periodically insert an undersampling level between the different layers of a ConvNet. This is done in order to progressively reduce the spatial dimension of each incoming activation map, the amount of parameters and computation in the network, and thus also to control over-fitting. By replacing a group of pixels with a single pixel that represents either the maximum value (Max pooling) or the average of that group of pixels (Average pooling). The pooling operation is however quite different.

The clustering operation works on small grid regions of different sizes in each layer, and produces another layer with the same depth. Unlike convolution operations that simultaneously use all feature maps in combination with a filter to produce a single feature value, pooling works independently on each feature map to produce another feature map. Figure 2.6 shows an example of a pooling operation with both maximum and average functions. The sliding window is 2 × 2 in size and moves 2 by 2 along the y-axis and x-axis.

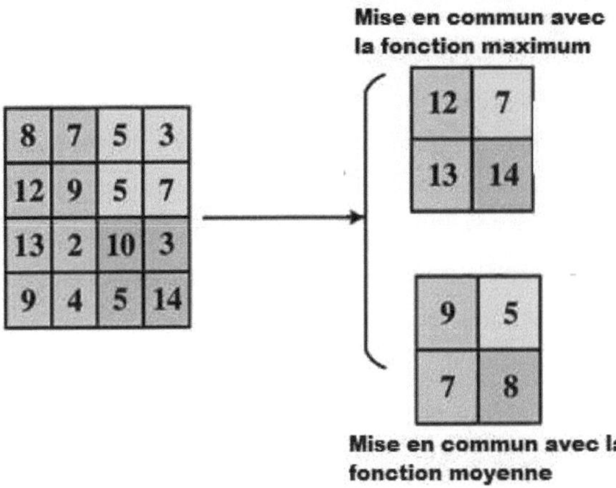

Figure 2.6: Example of possible pooling operations. The sliding window is 2 × 2 in size and moves 1 by 1 along the y-axis and x-axis.

D. Fully (completely) connected layers:

The fully connected layers, shown in Figure 2.7, like the other layers, can be added one after the other. In most cases, the fully connected layer is used to increase the computational power towards the end. Fully connected layers are sometimes used as the second to last layer of the CNN. They take the high level filtered images, they are identical to multilayer perceptrons (MLP).

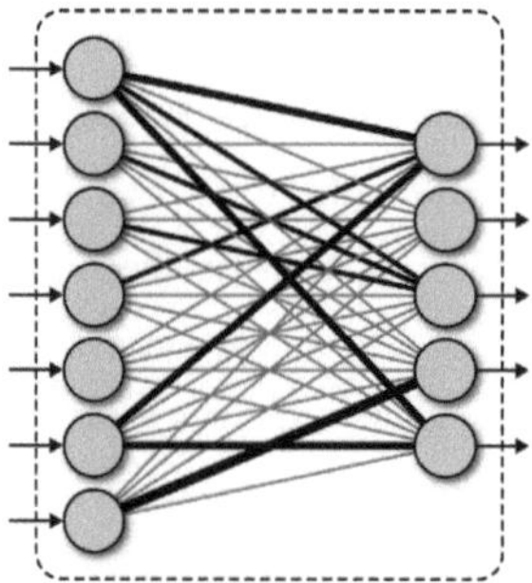

FIGURE 2.7 - Fully (completely) connected layers.

In this layer, each neuron is connected with all combinations of the outputs of the neurons in the previous layer which leads to a connection density that is very high. For example, if each of the two fully connected layers has 4096 hidden units, then the connections between them have more than 16 million weights (4096 x 4096). Similarly, the connections from the last space layer to the first fully connected layer will have a large number of parameters.

E. Batch Normalization

This layer is introduced in 2015 by researchers at Google [15]. It is usually located after fully connected layers or convolutional layers, and before nonlinearity. It alleviates the problems associated with internal covariance mismatch in feature maps by normalizing the layer's output distribution to zero mean unit variance. The use of this layer makes networks more robust to poor initialization, and it eliminates the need to use the dropout layer.

F. Dropout:

This is a popular [16] and simple method to prevent neural networks from overfitting during training, these layers ultimately improve generalization by randomly skipping a selectable percentage of their connections. Neurons that are "dropped" in this way do not contribute to propagation and do not participate in backpropagation. At test time, all neurons are used but their outputs are multiplied by the probability (typically 0.5 most used). This layer introduces regularization within the network, it is an extremely efficient regularization technique used to control the ability of neural networks to prevent overlearning. The disadvantage of this layer is that it approximately doubles the number of iterations required to converge. Figure 2.8 shows an example of a neural network after applying the dropout layer.

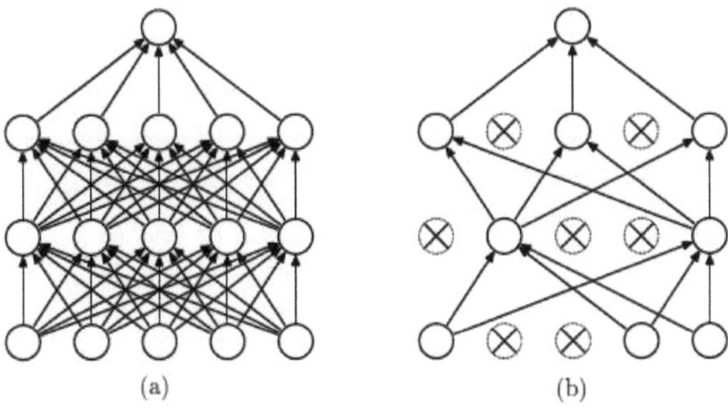

(a)   (b)

Figure 2.8- (a) Standard neural network; (b) After applying the "Dropout" layer.

G. Increase in data :

Data augmentation is a strategy that greatly increases the diversity of data available for model training without the need to collect new data, i.e., creating modified copies of each instance in a database. Data augmentation techniques such as cropping, filling and horizontal tilting are commonly used to train large neural networks.

2. Initialization of weights :

The performance of a CNN depends on how its parameters are initialized when it starts training. Typically, the weights of convolution networks must be initialized with small random numbers. A CNN can be considered as a function with learnable parameters and these parameters are often called weights and biases. They are initialized in different ways sometimes, using constant values such as 0 and 1, and sometimes with a normal distribution, sometimes with other techniques like xavier initialization.

A. Initialization with constants :

When initializing with constants, the weights of all neurons are usually set to zero or one. All neurons in each layer perform identical computations (all neurons learn the same thing) and thus provide the same output, so all neurons will compute the same gradients during back propagation [3].

B. Random initialization with a normal distribution

In this type of initialization, the weights are randomly distributed with a mean of zero and a standard deviation of one (1). This helps to give better accuracy. If a node in the hidden layer has inputs with these randomly distributed weights, the node will also be normally distributed around zero, but its derived variance and standard deviation will be greater than 1. Thus, the normal distribution of the hidden node will be wider than a normal distribution with a standard deviation of 1. A normal distribution is a distribution that depends on two parameters: its mean, denoted μ, and its standard deviation, denoted σ. The probability density of the normal distribution is given by:

$$f(x) = \frac{1}{\sigma\sqrt{2\pi}} e^{-\frac{(x-\mu)^2}{2\sigma^2}}$$

The notation for the normal distribution is :

$$X \sim \mathcal{N}(\mu, \sigma^2)$$

C. Xavier Initialization

In [17], the authors proposed in 2010 to adopt a uniform distribution for the initialization of weights, they called "Xavier Initialization" for the purpose of maintaining the variance of each neuron between layers. The variance of a given neuron depends on the variance of the weights in the previous layer. The Xavier initialization consists in reducing the variance from 1 to $\frac{1}{n}$ where n is the number of neurons. Variance reduction can produce better results [17].

D. He-and-al initialization:

This initialization method [18] is similar to the Xavier initialization with only one difference in the Xavier initialization factor which is multiplied by two [19]. This initialization of weights allows to reach a global minimum of the cost function with more efficiency and less time.

H. Recurrent neural network :

Recurrent neural networks are artificial neural networks with recurrent connections, so that activations can flow in a loop. They are made up of interconnected units (neurons) that interact non-linearly. These neural networks are closer to the real functioning of the nervous system, since information can travel in both directions, including from the deep layers to the first layers. Contrary to CNNs, Recurrent neural networks (RNNs) have been used for the analysis of temporal sequences, audio data, or texts. That is, when the data form a sequence and are not independent of each other. They are used in automatic recognition of speech or handwriting - more in automatic translation - or pattern recognition.

In sequential tasks, there is always a dependency of the current input data on the previous applied inputs. The task of RNNs is to find the relationship between the current input and the previous applied inputs. We note that in standard RNNs, this repeating module will have a very

simple structure, such as a single layer and its environment and all recurrent neural networks
have the form of a chain of repeating neural network modules.

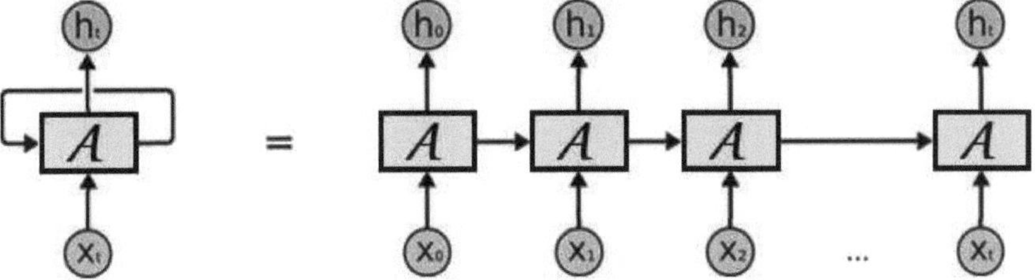

Figure 2.9 - Basic architecture of recurrent neural networks

Figure 2.9 shows the basic architecture of recurrent neural networks RNN unfolding into a
complete network. .

- X t is the input at time t. It is a vector of any size N.
- A is the hidden state at the time . It is the "memory" of the network. It is calculated
  according to the previous hidden state and the input at the current step. Represented by
  A_t where ω and U are weights for the input and the previous state value input, f is the
  nonlinearity applied to the sum to generate the final cell state.

$$A_t = f\left(\omega \times X_t + U \times A_{t-1}\right)$$

Recurrent neural networks of the Long Short-Term Memory (LSTM) type have proven
themselves in many applications. They work extremely well on a wide variety of problems and
are now widely used. Nevertheless, these approaches are designed to process only one input
sequence at a time. LSTMs were introduced in 1997 by Hochreiter et al [20]. LSTMs are
capable of remembering information for long periods of time is practically their default
behavior. They are explicitly designed to avoid the long-term dependency problem. Figure 2.10
shows the basic architecture of LSTM.

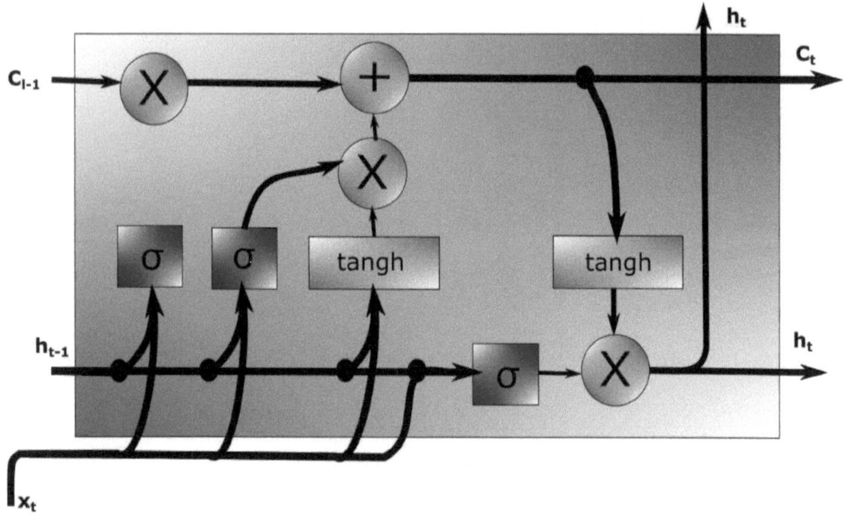

Figure 2.10 - Basic architecture of LSTM.

In addition, three "gates" are used to move information into or out of this memory:

- The input gate can allow the incoming signal to change the state of the memory cell or to block it.
- The output gate can allow the state of the memory cell to affect other neurons or prevent it.
- The forget gate can modulate the self-recurrent connection of the memory cell, allowing the cell to remember or forget its previous state, as needed (i.e., set to zero).
-

I. Advanced techniques to improve performance

Several techniques that are used to improve the performance of CNNs: increasing the depth; increasing the width; changing the convolution parameters [21-25] or pooling [26-33]; changing the activation function and reducing the number of parameters and resources. Note that depth and width refer to the capacity of CNNs, network depth means the number of layers, network width refers to the number of units in each layer. The number of feature maps (channels) in each layer can represent the width (breadth) of CNNs.

Image classification with convolutional neural networks.

In scientific research, a comparison between the different models proposed is required. This is what is required in the following where we will try to represent the different models of convolutional neural networks. We will first look at the different architectures that have won the first places in the annual ImageNet competition. We note that the software models of the implementation of convolutional neural networks consist in the use of physical media based on CPU or GPU. In this section, we will provide a view on the improvements and evolution made on the topologies of various deep CNN architectures (depth, width, multipath...) that are trained on the basis of ImageNET data. Various CNN models have been proposed since the remarkable performance of AlexNet on the ImageNet data base in 2012. These innovations were mainly due to the reorganization of different layers and the design of new blocks or else the exploitation of multipath. Table 3.1 summarizes the number of layers for each topology, including the convolution layer, the pooled, fully connected layer. We note that the activation function exploited by these different topologies is the ReLU function. The different architectures of deep neural networks proposed so far have mainly focused on improving accuracy. Although the accuracy figures have steadily increased, the resource utilization of the winning models has not been properly taken into account.

Model	Ref	Conv	Pool	FC
AlexNet	[10]	5	3	3
GoogleNet	[12]	7	14	1
VGG-16	[14]	13	5	3
VGG-19	[14]	16	5	3
ResNet-50	[15]	48	2	1
ResNet-101	[15]	99	2	1
ResNet-152	[15]	150	2	1

Table 3.1 - The number of layers for each model

Table 3.2 summarizes the number of Multiply Accumulate Operations (MACC) and the number of parameters for the different architecture where most of the MACC operations are consumed during the convolution layers.

Models	Ref	Conv	Conv	FC	FC	Total	Total
-	-	MACC	Param	MACC	Param	MACC	Param

AlexNet	[35]	666 M	2.33M	58.6M	58.6M	724M	62M
GoogleNet	[38]	1.58G	5.97M	1.02M	1.02M	1.58G	6.99M
VGG16	[37]	15.3G	14.7M	124M	124M	15.5G	138M
VGG19		19.5G	20M	124M	124M	19.6G	144M
ResNet50	[39]	3.86G	23.5M	2.05M	2.05M	3.86G	25.5M
ResNet101		7.57G	42.4M	2.05M	2.05M	7.57G	44.4M
ResNet-152		11.3G	58M	2.05M	2.05M	11.3G	60M

Figure 3.2- The number of MACCs and parameters for the different models}.

I. LeNeT :

LeNet[34] was one of the first convolutional neural networks. It is the model that promoted the development of deep learning. Moreover, it is described as the basis for other deep learning models. This old neural network architecture was designed in 1998 by Yann LeCun et al. The LeNet architecture, shown in Figure 3.1, was mainly used for digit recognition and applied to handwritten number recognition. In general, LeNet[34] refers to lenet-5 which is a simple convolutional neural network. The architecture consists of a total of 7 layers consisting of 2 sets of convolution layers and 2 sets of pooling layers which are followed by a flattening convolution layer. After that we have 2 fully connected dense layers and finally a softmax classifier.

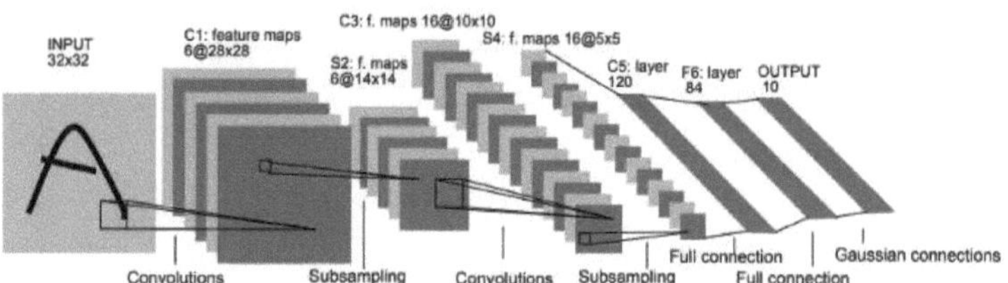

Figure 3.1 - The LeNet model [34].

If we take a standard incoming grayscale image (32x32) from the MNIST database that passes through the 6 filters of the first convolution layer having the size of the kernel (5x5) and with a step size of 1. The result of this convolution should lead to the change of dimension from (32x32x1) to (28x28x6) . The change of the number of channels from 1 channel to 6 channels is due to the application of 6 filters to the input image. In the second layer, a pooling layer is applied with a filter size of (2x2) and a step of 2. After this layer, the resulting image size will decrease to (14x14x6).

In the third layer, a convolution of 16 filters with a kernel size of (5x5) is applied with 16 feature maps. This convolution causes the image size to change from (14 x 14 x 6) to (10 x 10 x 16) . A pooling layer is applied in the fourth layer with the filter size (2x2) and a step size of

2. The resulting image, after this layer, will be of the dimension (5x5x16). The fifth layer represents a fully connected layer that has 120 neural units. Another fully connected layer of 84 neurons is applied in the sixth layer. The result is then passed through a sigmoidal activation function.

II.   AlexNET

Introduced by Alex Krizhevsky et al in 2012. It is described among the first convolutional neural networks for image classification. It to win the ILSVRC (ImageNet Large Scale Visual Recognition Challenge) in 2012, consisting of 5 convolutional layers followed by 3 fully connected layers. This architecture has significant importance in the new generation of CNNs and has opened a new line of research on CNNs. AlexNET[35] is the first architecture that used the non-saturating activation function ReLu (rectified linear unit) for the non-linear part, instead of a Tanh or Sigmoid function that are the standard for traditional neural networks. This function is placed after each convolutional and fully connected layer (FC). The use of this function improves the convergence rate by alleviating the problem of gradient vanishing. Figure 3.2 shows the architecture of the AlexNet model.

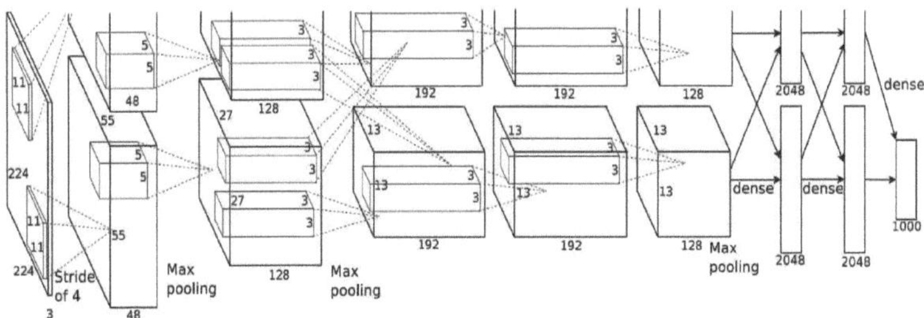

Figure 3.2 - The AlexNet model [35].

This architecture [35] uses five convolution layers with a local response normalization (LRN) layer after the first and second convolution layers and a MAX-pooling layer after the first, second, and fifth convolution layers. The advantage of this architecture is that it solved the overfitting problem by using a Dropout layer after each FC layer. This network required 62.4 million parameters and offered an ImageNet TOP 5 error rate equivalent to 16.4%. The AlexNet model was originally trained on a GTX 580 GPU with 3 GB of memory, and it was impossible to fit intermediate computations into this amount of space. Therefore, the network was partitioned onto two GPUs working together to create the training model with faster speed and memory sharing. It is noted that this model is trained with approximately 90 cycles using stochastic gradient descent (SGD) with :

- Momentum of 0.9.
- Weight reduction 0.0005.
- Lot size 128.
- The learning rate was initialized at 0.01 and reduced three times before the end where it divides by 10 when the validation error rate stopped improving with the current learning rate.

The weights are initialized from the zero mean Gaussian distribution with a standard deviation of 0.01. The biases conv1 and conv3 are initialized to 0, the other biases are initialized to 1. Figure 3.3 shows the 96 convolutional kernels of size $11 \times 11 \times 3$ learned by the first convolutional layer on the $224 \times 224 \times 3$ input images. The best 48 kernels were learned on GPU 1 while the last 48 kernels were learned on GPU 2.

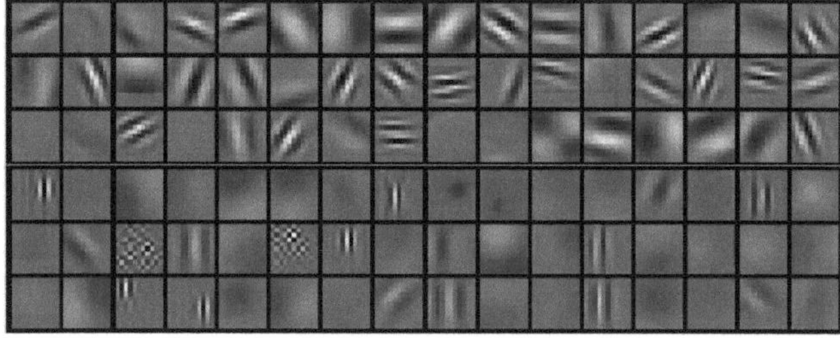

Figure 3.3- The 96 convolutional kernels of AlexNet [35].

The AlexNet model used two forms of data augmentation: the first form of data augmentation consists of generating image translations, horizontal reflections and randomly extracting patches of size 224×224 from the images of size 256×256 and training their model on these extracted patches. The second form of data augmentation is to modify the RGB channel intensities in the training images. More precisely, Principal Component Analysis (PCA) is performed on the set of RGB pixel values in the training set. Figure 3.4.(b) shows the resulting image after data configuration and data pre-processing for an incoming image (Figure 3.4.(a)).

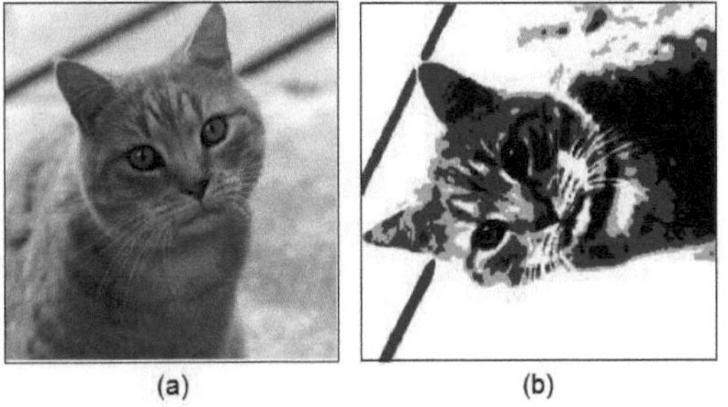

(a)   (b)

Figure 3.4- (a) incoming image, (b) the resulting image after data configuration and data pre-processing.

Figure 3.5 shows examples of the resulting images after the first convolution layer, note that the incoming image is the one in Figure 3.4.(a).

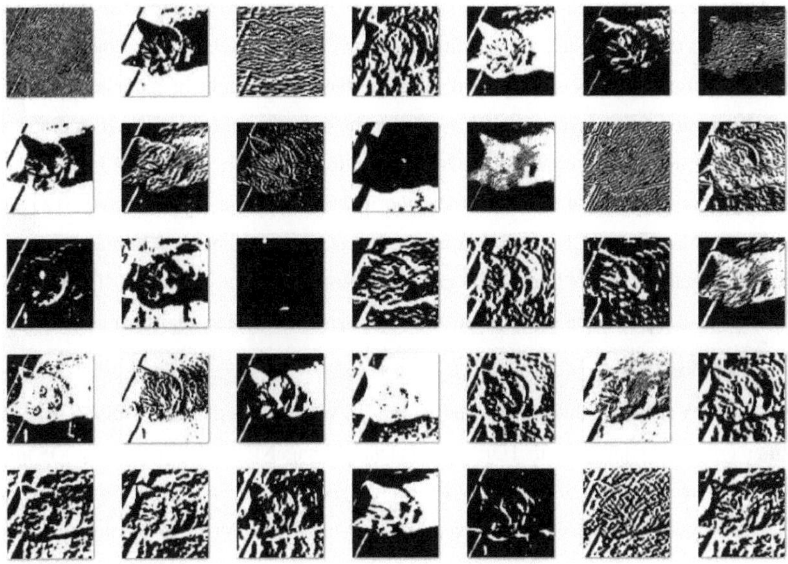

Figure 3.5- Examples of resulting images after the first convolution layer

III.   ZF NET :

In 2013, Zeiler and Fergus introduced a multi-layer deconvolutional neural network known as (ZFNET)[36], where they proposed a new visualization technique that gives a quantitative insight into the performance of the network. The idea of network activity visualization is to discover the contribution of different model layers to the performance by monitoring the learning pattern during training and use these results to extract the problem associated with the model. Experimental validation of this idea has been performed on AlexNet [35]. The ZfNet model is described in Figure 3.6

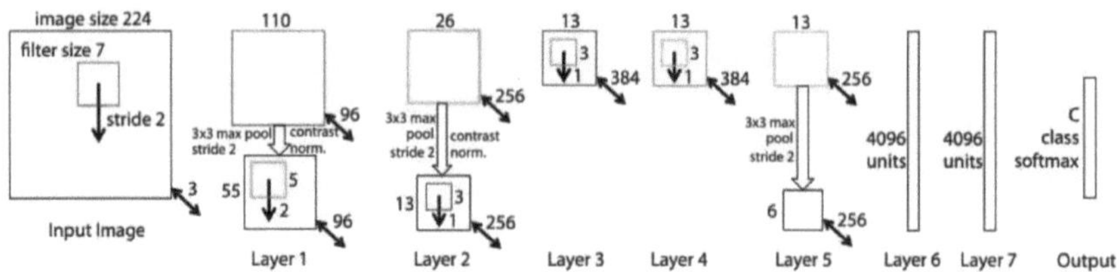

Figure 3.6 The ZFNet model [35].

The experimental results showed that only a few neurons were active, in contrast to the other neurons that were inactive in the first two layers of the model. Based on these results, the authors modified the CNN topology and optimized the CNN learning by reducing both the size of the filters in the first layer from (11 x 11) (Stride: 4) to (7 x 7) (Stride: 2). In addition, they increased the number of kernels in the 3rd convolution layer from 384 to 512 kernels, the 4th convolution layer from 384 to 1024 kernels, the 5th convolution layer from 256 to 512 kernels. This readjustment of the CNN topology resulted in an improvement in performance where this architecture achieved a Net Image error rate equivalent to 11.7% instead of 16.4% for [35]. The feature visualization can be exploited to extract design issues and to adjust parameters.

IV.   VGGNET

In 2014, the VggNeT architecture proposed in [37] won part of the 2014 ILSVRC challenge with an error rate equivalent to TOP 5 equivalent to ImageNET 7.4% and nearly 138 million parameters knowing that it uses fully connected layers. There are several topologies with different numbers of layers ranging from 11 to 19 layers (the VGG-16 topology has a very regular structure of 16 layers). Figure 3.7 shows the architecture of the VGG 16 model.

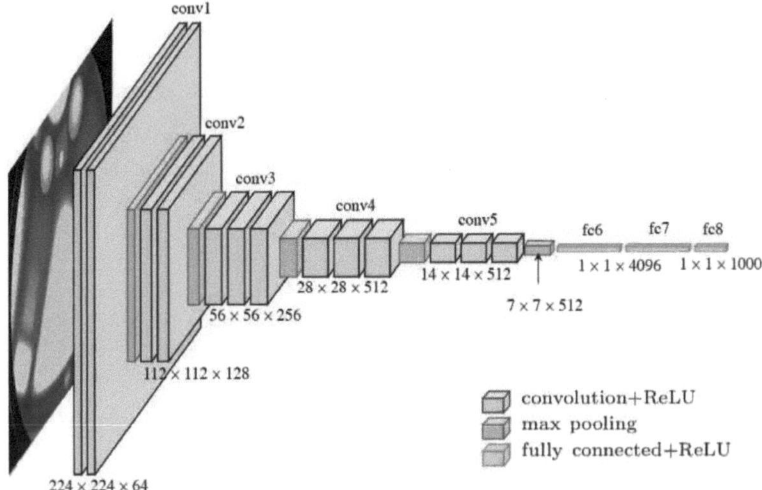

Figure 3.7- The VGG-16 model [37].

This architecture replaced the 7 x 7 convolution filters with a step size of 2 by three (3) convolution filters of size 3 x 3 (stride 1, pad 1). The goal was to keep almost the same resulting dimension of the 7 x 7 convolution while decreasing the number of parameters and increasing the number of activation functions used. In addition, this architecture uses a maximum sampling penny of size 2 x 2. VGG16 [37] is based on convolution filters of size 3 x 3 and the width of the network starts from 64 and increases by a factor of 2 every 2 convolution layers. We note that VGG [37] is the most expensive architecture both in terms of computational requirement and the number of convolution parameters from 384 to 1024 cores, the 5th convolution layer from 256 to 512 cores. This readjustment of the CNN topology resulted in improved performance. The training details are as follows:

- Algo. Optimization: SGD with a Momentum equivalent to 0.9.
- Weight decrease: 0.0001;
- Batch size: 256;
- Epochs: 74; 370K Iteration;
- The learning rate was initialized at 0.01 and reduces 3 times before the end where it divides by 10 when the validation error rate stopped improving.

V. GoogleNET :

Introduced by Christian Szegedy et al, it won the 2014-ILSVRC competition. GoogLeNet [38] contains 22 layers and achieves an ImageNet TOP 5 error rate equivalent to 6.67% while only requiring nearly 7 million parameters. These savings are achieved through a complex architecture that uses a sub-architecture in the network called "inception" modules. Figure 3.8 shows the architecture of the GoogleNet model.

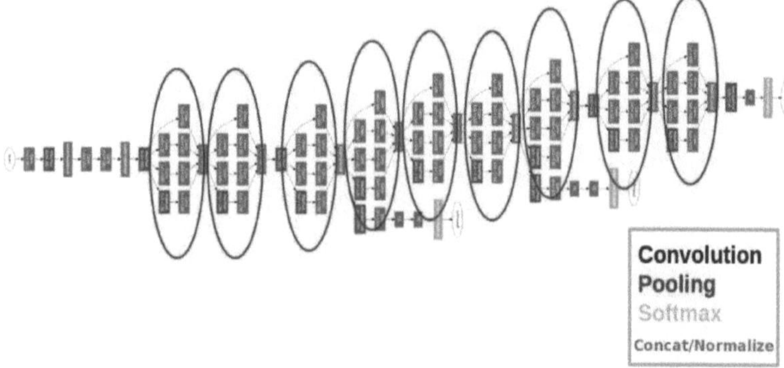

Figure 3.8- The GoogleNet model[38].

The main target of this architecture was to reduce the computational cost while offering high accuracy, this can be noticed from the 1×1 convolution layer before the 3x3 and 5x5 layers aiming to reduce the number of channels. The reduction of the channel size automatically causes a decrease in the number of parameters and MACC operations. In addition, a global average pooling layer is used at the last layer, instead of a fully connected layer in order to reduce the density of connections. Figure 3.9 shows the architecture of the inception module of the GoogleNet model.

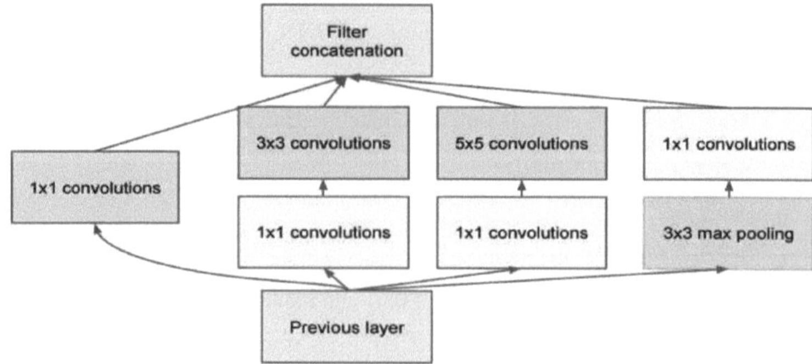

Figure 3.9 - GoogLeNet's "Inception" architecture[38].

The authors of the GoogleNET architecture [38] added auxiliary classifiers connected to intermediate layers until 'increasing the back-propagated gradient signal (Gradient back-propagation) and accelerating the convergence rate. At test time, these auxiliary networks are discarded and at training time, the losses of the auxiliary classifiers are balanced to 0.3. The training details are as follows:

- Algo. Optimization: SGD with a Momentum equivalent to 0.9.
- Epochs: 600K iteration;
- The learning rate decreases to 4% every 8 periods.

VI.   ResNET :

As everyone knows, after the invention of VGGNet [37] and GoogleNet [38], it was thought that as the architecture stacks more layers the accuracy becomes better. But deeper neural networks with more than 20 layers generally are more difficult to train hence training very deep networks requires slow training time and creates several problems such as vanishing gradients in back propagation.

In [39] He et al, presented a "Residual Block" to facilitate the formation of networks considerably deeper than [37] and other networks by including a link around every two convolutional layers, adding both the diverted original and its function. This solution is called "Residual Block". Each "residual block" incorporates 2 convolutional layers (3 x 3). Figure 3.10 shows the residual function.

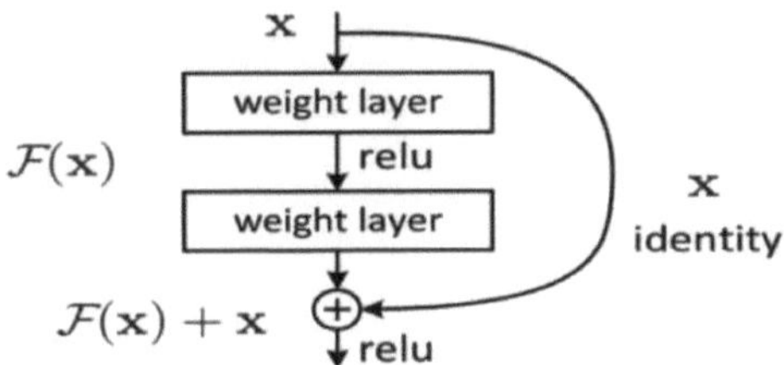

Figure 3.10 - The residual function [39].

ResNet [39] was proposed by He et al and offers a result that won 1st place in the ILSVRC 2015 classification task, for a residual network with a maximum depth (152 layers) almost 8 times deeper than [37] the error equals 3.57%. We note that even with a deeper architecture, ResNet [39] has a lower computational complexity than [37]. This architecture is similar to

GoogLeNet [38] in the use of a global average sub-sampling followed by the classification layer, so this architecture does not use the dropout layer.

VII.   ResNet Large :

Wide ResNet [40] is proposed by Zagoruyko and Komodakis. This work [40] supports the principle: "Residual Blocks are the important factors in the architecture and not the depth which is an additional effect". This principle is necessarily validated when we notice that a 50-layer wide Large-ResNet outperforms the original 152-layer ResNet on ImageNet.

WRN [40] exploited a Wide Residual Block that integrates (L x K) filters instead of L filters like [39], where k is defined as an additional factor, which controls the width of the network. Wide ResNet [39] showed that the use of a wider Residual Block represents an effective solution to improve the accuracy that deepening residual networks. The training details are as follows:

- Algo. Optimization: SGD with a Momentum equivalent to 0.9.
- Weight decrease : 0.0001;
- Batch size: 256
- Epochs: 600K iteration;
- The learning rate has been initialized to 0.1 and is divided by 10 when the validation error rate has stopped improving

VIII.   ResNEXT

In [41], the ResNeXt model is similar in spirit to the "Inception" module of GoogLENET[20]. ResNeXt is illustrated in Figure 3.11 It is another architecture that is proposed by the designers of ResNET[39] in 2017 where they have increased the width of the residual block through multiple parallel paths. This strategy unveils a new dimension, called "cardinality".

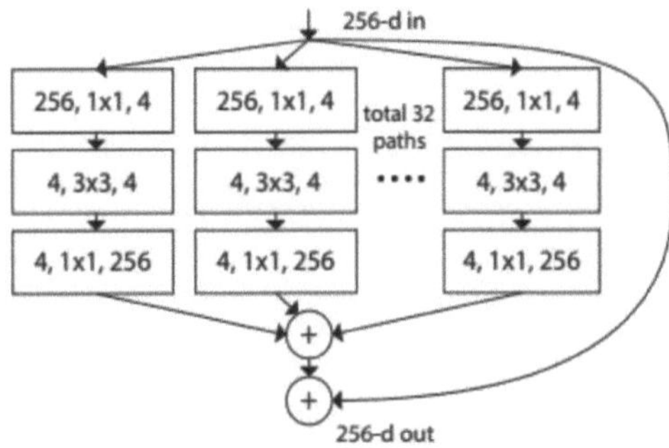

Figure 3.11- The ResNeXt model. [41]

The test results showed that increasing cardinality is a solution to improve the classification accuracy. Moreover, it is more efficient than deepening the networks. This architecture used a deep homogeneous topology of VGGNET [37] and the simplified heterogeneous architecture of GoogleNet. ResNeXt [41] achieved an error rate of 3.03%(Top-5) and wins 2nd place in the ILSVRC 2016 classification task. The training details are as follows:

- Algo. Optimization: SGD with a Momentum equivalent to 0.9;
- Weight decrease : 0.0001;
- Batch size: 256 on 8 GPUs (32 per GPU).
- Periods: 300 ;
- The basic learning rate is 0.1. It divides by 10 three times. The learning rate (CIFAR) decreased by 0.2 at 150 and 225 epochs.

IX.   **DenseNeT** :

It is widely known that convolution networks can be much deeper, and more efficient during training if they contain shorter connections between the layers close to the input and those close to the classification layer. Convinced of this principle, the authors proposed in [43] a "DenseNET" architecture which is based on dense blocks where each layer is connected to all other layers in a feedforward mode. Thus, feature maps from all previous layers were used as inputs to all subsequent layers. This reinforces the propagation of features, and encourages their reuse. This network establishes $\frac{L \times (L+1)}{2}$ direct connections since it connects every layer to every other layer in a retroactive way. Whereas traditional convolutional networks with L layers have "L" connections.

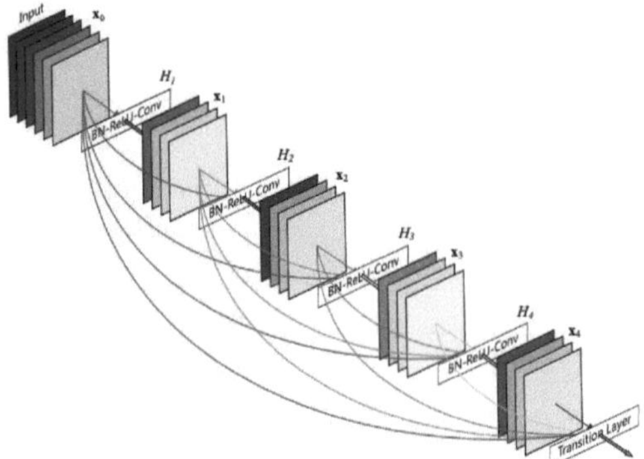

Figure 3.12- DenseNet architecture [43]

DenseNET [43], illustrated in Figure 3.12, offers several advantages such as mitigating the problem of vanishing gradients, encouraging the propagation and reuse of feature maps. This leads to the reduction of the exploited parameters.

X.   Network in Network (NiN)

The "Network In Network" model [44], shown in Figure 3.13, consists of several "MLPconv" layers that are stacked in a successive manner. The MLPconv layer consists of a linear convolution layer and a two-layer MLP with a ReLU used as an activation function.

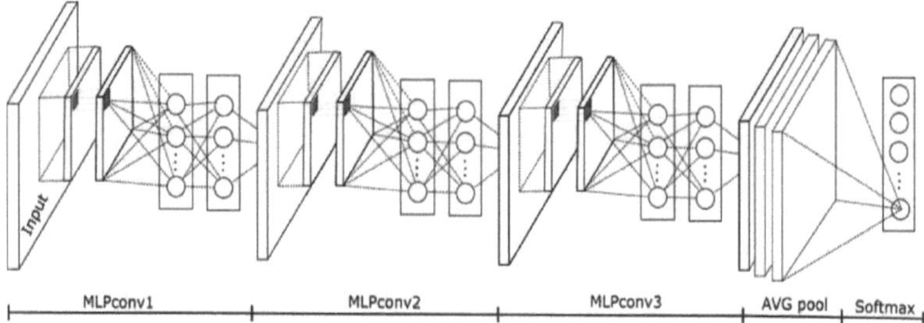

Figure 3.13- Network In Network[44].

In the NiN model [44], a global average pooling layer is used instead of the fully connected layers that are traditionally used in CNNs. This solution allows to generate a feature map for each category. The average of each feature map and the resulting vector will be fed directly into the Softmax layer. The advantage of this technique is that there is no parameter to optimize, which avoids over-fitting in this layer. In terms of accuracy, the NIN model [44] obtained an error rate equivalent to 10.41% using the dropout layer and without data augmentation. Using the data augmentation layer (horizontal translation and flipping), the NIN model [44] obtained an equivalent error rate of 8.81%. For the training of the NiN model, an adaptation of the training procedure used by AlexNet [35] is exploited. The network is trained using mini-lots of size 128. The training process starts from the initial weights and when the accuracy on the training set stops improving, the training rate is reduced. This procedure is repeated once so that the final learning rate is one percent of the initial value.

XI. Network In Network deep:

The deep-dimensional network (DNiN) [45], shown in Figure 3.14, represents an important extension of NIN [44]. In this work, the authors used two filters with a very small size (3 × 3, 1, 1) instead of the size filter (5x5, 1, 2) in NIN [44]. This type of filter is defined as the smallest size to capture the notion of left / right, top / bottom, center.

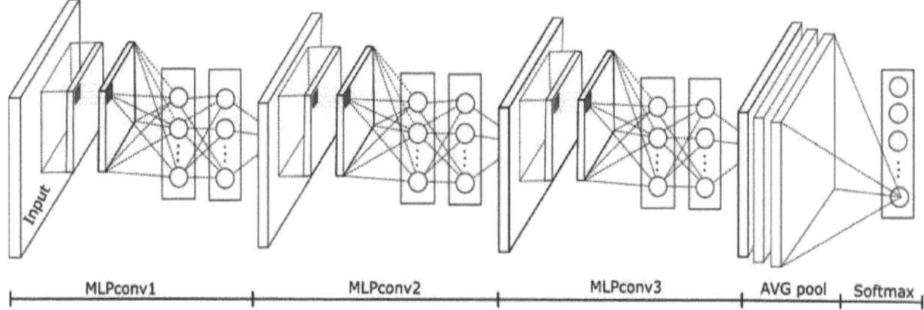

Figure. 3.14 "Deep Network in Network" [45].

In addition, the authors used the linear exponential unit (eLU) to solve the leakage gradient problem that can occur when using the ReLU function and to speed up the learning process. In this model, a batch normalization layer is applied to reduce eLU saturation and the "Dropout" layer is applied to avoid overlearning. The experimental results on CIFAR-10 database show that the reduction of filter size shows an improvement in the recognition accuracy of the model.

XII. SENets

The SENets (Squeeze-and-Excitation Networks) model, shown in Figure 3.15 and described in [46], introduces a building block for CNNs that improves channel interdependencies with almost no computational cost. It adds a feature recalibration module to networks such as ResNets, allowing them to learn to reweight feature maps instead of simply copying them between layers. They can be easily added to existing architectures. The main idea of this work is to have parameters added to each channel of a convolutional block so that the network can adaptively adjust the weight of each feature map.

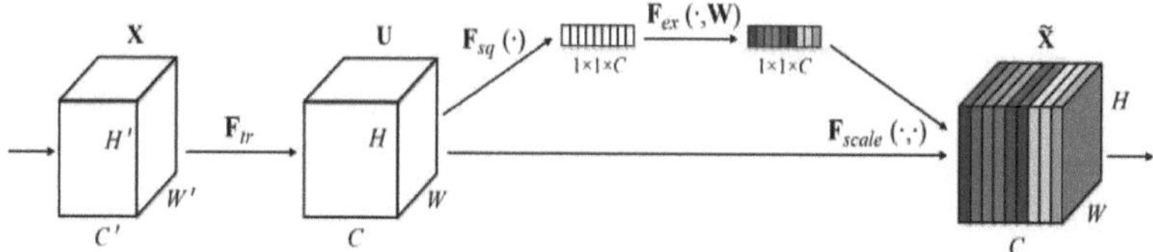

Figure 3.15 - SENet [46].

XIII. SqueezeNet:

The SqueezeNet model, shown in Figure 3.16, is described in [47], it used an automatic pattern search approach by exploring different possible combinations of 3x3 and 1x1 filters on each layer. It achieved equivalent accuracy to AlexNet with a more efficient model, to which several compression methods were then applied.

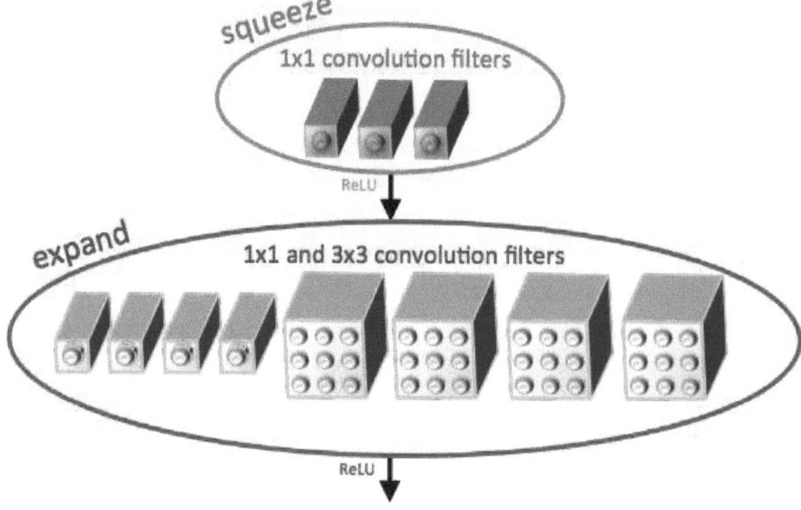

Figure 3.16 - SqueezeNet [47].

XIV. FractalNet:

FractalNet, shown in Figure 3.17 and described in [48], argues that residual connections may not be a necessary factor and proposes an architecture with multiple branches in each layer

38

that can be independently disabled during learning and achieve similar accuracy to ResNets for similar parameter budgets. Since all branches are fixed and used at inference time, the implementation could also be more efficient than ResNets.

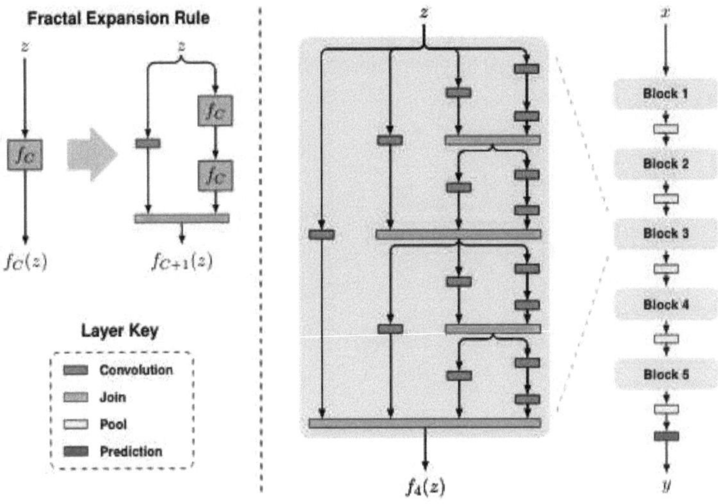

Figure 3.17: FractalNet [48].

XV.   MobileNets

MobileNets, described in [49], use Depthwise Separable Convolution, i.e., each channel is convolved separately with a filter, and the resulting feature maps are then recombined in a learnable way, instead of a simple summation across channels, as done in traditional CNNs . This achieves good tradeoffs between model size and complexity. Figure 3.18 shows the depthwise separable convolution r(Depthwise Separable Convolution).

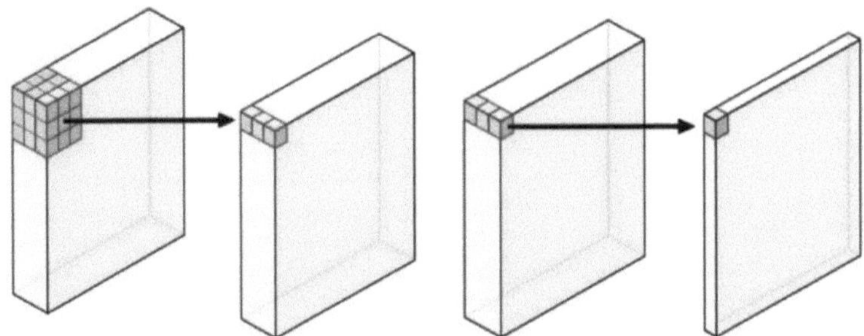

Figure 3.18 - Depthwise Separable Convolution [49].

XVI. ShuffleNets

This model is described in [50], it proposes to use pointwise group convolutions to reduce the computational complexity of 1×1 convolutions as well as a channel suffle operation to overcome some side effects brought by group convolutions. Figure 3.19 shows the ShuffleNet unit with deep convolution.

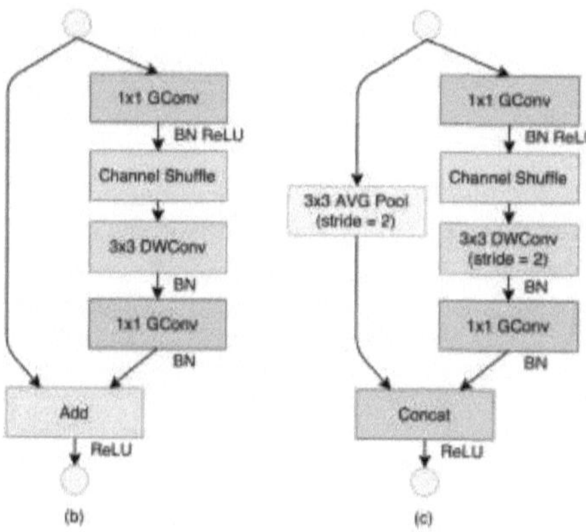

Figure 3.19 - The **ShuffleNet** unit with deep convolution [50] .

Techniques for developing and accelerating deep learning.

Convolutional neural networks are at the heart of many deep learning algorithms and have been used successfully to solve classification problems and many other applications. However, they require intensive processor operations and memory bandwidth[3] that prevent general-purpose processors from achieving desired performance levels. Therefore, hardware accelerators that use ASICs, FPGAs and GPUs have been used to improve the throughput of CNNs. In particular, FPGAs have been adopted and widely exploited to accelerate the implementation of deep learning models due to their energy efficiency as well as their ability to maximize parallelism. The following chapter focuses on the optimization and compression methods and techniques used in FPGA-based deep learning acceleration and to improve the acceleration performance.

I.   FPGA (field programmable gate array)

FPGAs are programmable devices that provide a flexible platform for implementing custom hardware functionality at low development cost. They consist primarily of a set of programmable logic cells, called configurable logic blocks (CLBs). CLBs are the heart of the FPGA. They are cells made up of programmable logic elements including flip-flops (registers), LUTs (Look-Up Tables), multiplexers and logic gates arranged in a matrix form, and a hierarchy of programmable interconnects allowing the blocks to be "wired together", and a set of programmable input and output cells around the board.

FPGAs incorporate a rich set of integrated components such as digital signal processing (DSP) blocks that are used to perform arithmetic operations (multiplication, accumulation, etc.), BRAMs, LUTs, flip-flops (FFs), a clock management unit, I/Os (input/output cells that allow the FPGA to be interfaced with the external environment), etc. Figure 4.1 shows the basic structure of the FPGA.

---

[3] Memory bandwidth is defined as the speed at which data can be read from or stored in a memory by a processor.

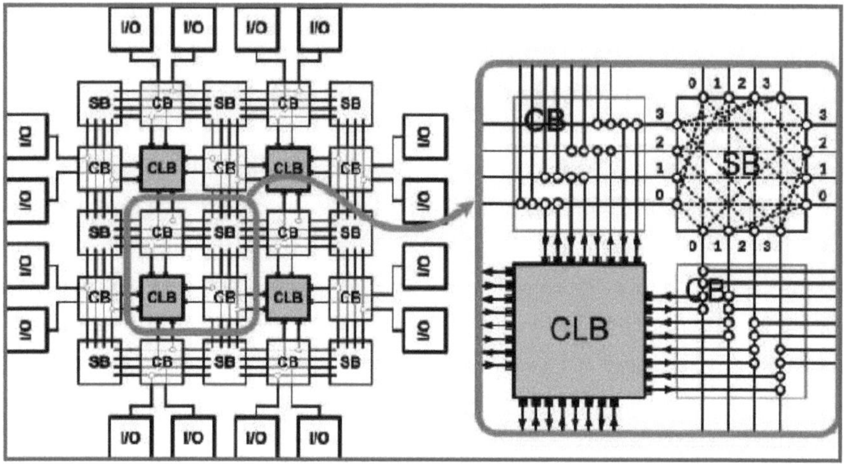

Figure 4.1 - the basic structure of the FPGA.

FPGAs are capable of programming logic, computation, processing, image and signal functions. In addition, they are widely regarded as accelerators for computationally intensive applications, enabling associative operations such as data streaming and broadcasting, highly parallel memory access, use of standard hardware structures such as buffers (FIFO), stacks and priority queues...

FPGAs have the advantage of maximizing performance per watt of power consumption, which reduces costs for large-scale operations [51]. This makes them an excellent choice as accelerators for embedded electronic systems and battery operated devices. FPGAs also play a remarkable role in embedded system development due to their ability to start software system (SW) development simultaneously with hardware (HW). FPGAs have been widely exploited for deep learning acceleration given their flexibility in implementing architecture with a high degree of parallelism allowing high execution speeds [51].

The FPGA configuration is usually specified using a hardware description language (HDL), similar to that used for ASICs. The adoption of software-level programming models such as the OpenCL standard [52, 53] in FPGA tools has made them more attractive to use for deep learning [54], [55]. FPGAs offer an advantage for deep learning algorithms since they can create custom hardware circuits that are deeply pipelined [51]. FPGAs also have the capability of partial dynamic configuration, which allows a portion of the FPGA to be reconfigured while the rest is in use. This could be a potential advantage for deep learning methods where the next layer could be reconfigured while the current layer is in use.

II.   Hardware acceleration of deep learning networks :

CNNs have become useful for embedded applications in different domains (autonomous robots, automobiles) that require real-time execution and high accuracy. CNN algorithms are very computationally expensive due to the high temporary consumption of convolution operations [56]. Moreover, the increasing evolution of deep learning networks generally results in more complex CNN structures as well as deep CNN models. Thus, millions of parameters and billions of operations, as well as substantial computational resources are required to train and evaluate the resulting large-scale CNNs [55, 57]. These requirements represent a computational challenge for general purpose processors (GPPs). Therefore, hardware accelerators such as ASICs, FPGAs, and GPUs have been used to improve CNN throughput. In practice, training is performed offline using the backpropagation process [58]. Then, these offline trained CNNs are exploited to perform recognition and classification tasks using the feed-forward process [139]. The execution of CNN inference requires much less computation. In addition to GPUs, inference can be executed on CPUs, TPUs, FPGAs and ASICs. Running a complete training process on an FPGA is a very challenging task, due to the limited resources of FPGAs and the complexity of back propagation[58]. For this purpose, the training and execution is performed offline, using GPUs or CPUs. GPUs are the most widely exploited hardware accelerators for improving training and classification processes in CNNs [59]. Due to their memory bandwidth and high throughput, as they are very efficient in floating point matrix operations [60-62], CNNs need to perform Multiplication-Accumulation (MACC) operations with floating point numbers. However, GPU accelerators consume a large amount of power, which makes them incompatible with all real-time, low-power embedded systems or in battery-powered devices. In addition, GPUs derive their performance from their ability to process a large batch of images in parallel. For some applications, such as a video stream, the input frames must be processed frame by frame because the latency of the output of each frame is critical to the performance of the application.   In contrast, CPUs can execute instructions to perform the arithmetic calculations required for inference and training, but they do not always have the speed (compared to GPUs) to perform the amount of calculations required for CNNs, although they can use SIMD instructions and multiple cores to perform multiple arithmetic calculations in parallel. TPUs are ASICs designed primarily to perform neural network inference and improve the cost-performance of GPUs [56]. FPGAs and ASICs have relatively limited memory, I/O bandwidths and computational resources compared to GPUs. They can achieve at least moderate performance with lower power consumption [63]. The throughput of the ASIC design can be improved by customizing the memory hierarchy and allocating dedicated resources [64]. However, the development cycle, cost and flexibility are not satisfactory in ASIC acceleration of deep learning networks [65, 66]. As an alternative, FPGA-based accelerators are currently used to provide high throughput at a reasonable price with low power consumption and reconfigurability [67,68].

III.   Challenges of implementing FPGA-based DNNs.

Implementing deep learning networks and, in particular, CNNs on FPGAs presents a number of challenges, including the requirement for a large amount of storage, external memory bandwidth, and computational resources on the order of billions of operations per second [69]. Storage of a large amount of the weights is not supported by existing commercial FPGAs, therefore, the weights must be stored on an external storage tool and transferred to the FPGA during computation. For example, the AlexNet model requires 250 MB of memory to store the weights based on a 32-bit floating point representation [70]. The problem gets worse and exacerbated with more complex models such as the VGG-16 model which needs, for example, 138 million weights and requires more than 30 GOPS [71]. Current trends in CNN implementation are moving towards compressing the entire CNN model with bit quantization of the data [72].

As technology advances, FPGAs continue to grow in size and capability. Certain mechanisms are essential to meet the requirements for efficient implementation of deep learning networks. Solving hardware resource limitations requires the reuse of computational resources and the storage of partial results in internal memories, the design and implementation of flexible reconfigurable and reusable architectures is one of the best solutions. Data transfer and computational resource utilization are significantly influenced by the order of operations and the selection of parallelism in the implementation of CNNs on FPGAs. A significant reduction in external memory access and internal buffer size can be achieved by careful operation scheduling. In addition, the number of external memory accesses can be reduced by using on-chip memory and exploiting data reuse. The bandwidth requirements of external memory can also be reduced by using reduced precision to represent weights, which also results in better energy efficiency.

IV.   The flow of implementation and development:

The implementation and development of the CNN on FPGAs is usually done using high-level programming languages such as C+ and the Vivado HLS (High Level Synthesis) tool or using the VHDL description language and development environments such as Vivado (Xilinx). In addition, the OpenCL language, which is used to program GPUs, can also be used as an input language for implementation on Intel-Altera and Xilinx FPGAs (Vivado SDAccel). The use of deep learning development environments simplifies the implementation. As a result, many existing development environments (frameworks) are now extended to use GPUs, such as TensorFlow, Caffe . In the remainder of this section, we provide more details on the development environments.

1. High level synthesis tools

Although hardware description languages (HDLs) have the advantage of providing the best results in terms of performance (resources, computational throughput), in FPGA design, they require a good knowledge of digital circuit design and hardware architectures. In the case of complex algorithms such as CNN, the transcription task becomes more difficult due to the large number of computations and the high variability of workloads. As a result, the transcription of complex algorithms into hardware descriptions is a task requiring considerable development time. In response to these productivity issues, significant research efforts are devoted to the development of high-level synthesis (HLS) tools. In this context, two popular "C-like" HLS tools have been introduced: Vivado HLS and OpenCL for FPGAs.

2. Vivado HLS

Vivado HLS is part of the Vivado Design Suite, which is a software suite produced by Xilinx for HDL design synthesis and analysis. It replaces Xilinx ISE with additional functionality for system-on-chip development and high-level synthesis. In addition, it provides a programming environment similar to those available to software developers. The tool relies on pre-compile directives or *pragmas* to generate the RTL description of a design. Note that during the transcription process, the tool generates representations in SystemC, VHDL and Verilog, and it is possible that a component designed by HLS can be generated separately and integrated into a project.

3. Open Computing Language (OpenCL)

OpenCL is an open-source framework for parallel programming on heterogeneous architectures. Programs written in OpenCL can be executed on CPUs, GPUs and FPGAs. It is a combination of an application programming interface (API) and a programming language derived from C. OpenCL not only compiles C code into an RTL description, but also handles interfacing with external memory and communication between the host CPU and the FPGA core.

4. Development environment

CNN models can be designed in development environments (frameworks) such as Caffe [73], Theano [74], Tensorflow [75] etc... The frameworks are designed to use a processor or a graphical processor for training and/or inference. Frameworks simplify development and implementation. They provide functionality to create layers for models; perform model training and perform inference of models without backpropagation. TensorFlow [75] is an open source machine learning tool that is developed by Google. It is one of the most widely used AI tools in the field of machine learning. TensorFlow [75] provides stable Python and C++ APIs. Caffe designs model layers and configurations: kernel; padding; stride (etc.) in special files named (.prototxt). It stores the parameters in a Caffe model file [73]. The hyper-parameters (base learning rate, optimization algorithm, learning rate decay) are used to define how fast and aggressively the model changes the parameters of the hidden and output layers. The hyper-parameters used are defined in a configuration file (.prototxt). The configuration files along with the model description can be passed as parameters to the main scripts to begin training or inference. The model design and hyper-parameters should be described in python scripts for Theano [74] and Tensorflow [75] using low-level functions specific to the development environment. The Keras application programming interface can be used to simplify the use of the Tensorflow low-level functions. Unfortunately, none of the development environments offer solutions that can be implemented directly on an FPGA.

6. Different types of parallelism: Granularity of implementation

Due to the large number of computations required, hardware integration of real-time CNNs is a challenge, especially if we are aiming for low-power embedded applications. There are several characteristics of most learning models and applications that make them generally well suited for parallelization using hardware accelerators. We discriminate three types of methods for parallelizing and/or distributing computation that we can formalize according to the type of granularity of the parallelism: data parallelism, model parallelism, and pipeline parallelism. Data Distribution Parallelism or Data Parallelism consists of distributing data over several threads, while using the same model, running on several machines. This approach can be used if the data is too large to be stored and processed on a single machine or to speed up learning. The advantage of model parallelism is that it speeds up the backpropagation of the deep neural network and the speed of inference. This type of parallelism also requires a lot of communication between the agents, since it is usually necessary to have the activation of the neurons of an intermediate layer to calculate the next one. To share the neural network on different machines, there are several possible granularities: First, it is possible to split the different layers of the neural network on the available agents. This solution allows pipelining the execution of the network when several data are computed one after the other. Then, and with a finer level of granularity, it is possible to distribute the neurons present on each layer between the available agents according to several dimensions: height, width or channels. Finally and even more finely, it is possible to have each matrix operation of the neural network shared independently. A layer of neurons can require several different operations, which allows the load to be distributed even more finely on each agent. In addition to data parallelism, and in model parallelism, when a model is too large to fit on one machine to support the learning process, one option is to distribute/share it across multiple machines.

V.   Deep compression" techniques:

Current trends in hardware accelerators in deep learning use "deep compression" techniques with the goal of efficient inference. The use of CNN model compression methods provides a trade-off between throughput and accuracy. While it helps to reduce computational and storage complexity, it can sacrifice model accuracy. Deep compression mainly involves pruning and quantization techniques of the network. Compression reduces the number of parameters as well as the computational effort and inference time. The computational complexity can also be reduced by exploiting Winograd. We note that deep neural networks are compressed before their inference acceleration in hardware targets. there are several approaches that will be discussed in this subsection.

1. Quantification

The quantization method adopts reduced precision as well as smaller data representations of weights and activations. This is done to mainly reduce hardware cost, memory accesses and increase parallel executions. Quantization to the desired number of bits depends solely on the application requirements and precision. Therefore, it is difficult to speculate on an optimal precision that is sufficient for all purposes.

Generally, multiple accuracies such as INT4 [4], INT8, FP16, FP32 are used for different applications. Currently, INT8 data type is widely used for inference while FP16, FP32 data

---

[4] INT4: integer with a size (in bytes) equivalent to 4.

types are exploited for training. The use of INT 16 based operations saves about 30\% of FPGA resources. Many designs adopt multiplication with less than or equal to INT 16, which can result in a large underutilization of the DSP. It is noted that operations with 32 FP data consume similar resources as operations with INT 32. The 16-bit operations offer a $14\times$ reduction in hardware operators compared to the 32-bit operation. For 4-bit operations, the reduction rate becomes $54\times$. Quantization techniques are applied for weights and for activations. Normally, the activation quantization has a significant impact on the accuracy than that of the weight quantization. To improve the accuracy, the quantized parameters are refined by recurrent recycling with the original data set. Sometimes, to avoid recycling, a smaller floating point precision is exploited. Recent work attempts to replace the fixed-point representation. On the one hand, using fewer bits for each neuron or weight helps to reduce the bandwidth and storage requirements of the processing system of a CNN model. On the other hand, the use of a simplified representation reduces, as mentioned earlier, the hardware cost of each operation. Two types of quantization techniques are widely known: linear quantization [51, 52, 53, 54] and nonlinear quantization [55, 56, 57].

Non-linear quantization independently assigns values to different binary codes. The translation of a non-linear quantized code into its corresponding value is therefore a look-up table.

In [76], Podili et al used 32-bit fixed-point units for the proposed system. The work [77, 78, 79, 80, 81,] used 16-bit fixed-point units. Other approaches have suggested hybrid representations. In [82] 12-bit fixed weight neurons and 16-bit fixed point neurons are exploited. Indeed, in [83], the authors exploited a floating point representation for the CNN weights and a fixed point representation for the activation functions.

2. Pruning/ pruning :

The pruning mechanism was inspired to remove redundant parameters from neural networks that do not contribute significantly to the outputs (especially zero weights and activations). This procedure can be applied to all layers as a whole or to a specific layer of CNN. This facilitates the executions, which improves the computational performance. Pruning can be implemented using ranking methods. The rankings are usually based on the L1 / L2 weight standard, average activation, etc., taking into account the individual layers. The alternative technique used is iterative pruning, which is applied to the entire network. In this process, CNN is refined until the pruning objective is achieved. For example, the pruning goal could be in terms of defined model size or runtime performance. Figure 4.2 shows an example of neural network compression.

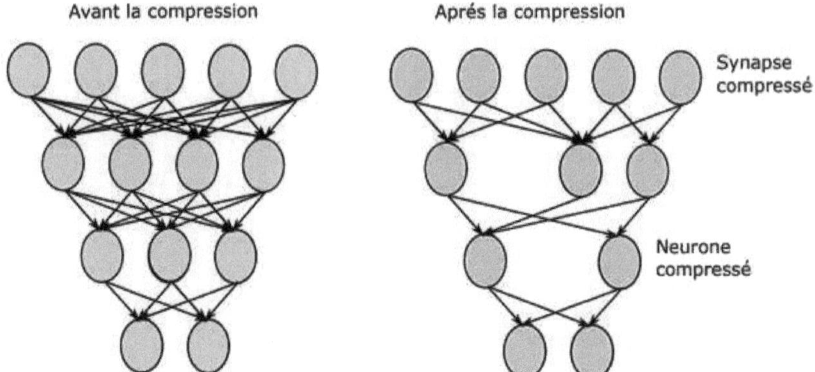

Figure 4.2 The compression of a neural network.

In other words, we can say that, pruning is another type of methods exploited to reduce the number of weights. This type of method directly removes the zeros in the weights or removes those with small absolute values. The problem with pruning is to reduce the weight to zero while maintaining the accuracy of the model. One solution is to remove the weights after training [85]. Another solution is to remove zero weights during training [84].

3. Transformation of Winograd :

The Winograd transformation is an optimized convolution algorithm that reduces the number of multiplier units and increases the functional throughput. These methods work well on 3x3 convolutions. In the Winograd transformation, the inputs and kernel parameters are transformed. The transformed values are subjected to elementwise multiplication, and then the results are converted back to the convolved output.

Suppose we have an input image f of size (4) and a filter g of size *(3)*.

$$f = [d_0\ d_1\ d_2\ d_3], \qquad g = [g_0\quad g_1\quad g_2]$$

Then, using the technique that rearranges the matrix blocks into columns (transposition), we convert the input image (matrix) into :

$$f = \begin{bmatrix} d_0\ d_1\ d_2 \\ d_1\ d_2\ d_3 \end{bmatrix}, g = \begin{bmatrix} d_0 \\ d_1 \\ d_2 \end{bmatrix}$$

We can now perform matrix multiplication using BLAS libraries such as CuBLAS (GPU) or Intel MKL (CPU) which are highly optimized for matrix multiplication.

For the Winograd calculation, instead of doing the scalar product, we calculate the resulting matrix using these formulas.

$$result = \begin{bmatrix} d_0 & d_1 & d_2 \\ d_1 & d_2 & d_3 \end{bmatrix} \begin{bmatrix} d_0 \\ d_1 \\ d_2 \end{bmatrix} = \begin{bmatrix} m_1 + & m_2 + & m_3 \\ m_2 + & m_3 + & m_4 \end{bmatrix}$$

Knowing that :

- $m_1 = (d_0 - d_2)g_0$

- $m_2 = (d_1 + d_2)\frac{g_0 + g_1 + g_2}{2}$

- $m_3 = (d_2 - d_1)\frac{g_0 - g_1 + g_2}{2}$

- $m_4 = (d_1 - d_3)g_2$

In this way, we can find the values of *m1, m2, m3, m4* . Then we can use them to calculate the convolution instead of the dot product of the matrices.

For $\frac{g_0 - g_1 + g_2}{2}$ and $\frac{g_0 + g_1 + g_2}{2}$, we can compute them once before convolution during training and can be saved during inference. Now we will need 4 ADD operations and 4 MUL operations to calculate the values of m1, m2, m3, m4 and 4 ADD operations in calculating the *result* using the calculated values of m1, m2, m3, m4. By doing normal scalar products, we would do 6 MUL operations instead of 4. This would reduce the computationally expensive MUL operations by a factor of 1.5x, which is very significant.

Now consider, for example, an input image of 4x4xC, kernel of 3x3xC. To compute the output feature map (2x2x3) the required functional units (MAC) include: the direct convolution requires (4x3x3xC) functional units 36xC. While, the Winograd expects 16 functional units x C. Therefore, it improves the performance by a factor of 2.25x.

4. Entropy encoding :

This is a well-known lossless compression method based on the idea of representing more common symbols by shorter codes. This type of compression has been applied to CNNs and combined with pruning in various works such as [86]. This type of compression requires specialized hardware in order to decode the weights and offer a lower compression gain than pruning.

VI.   Evaluation metrics:

Models implemented on FPGAs can be evaluated using certain metrics. Many metrics are available to highlight the strengths and weaknesses of the results produced by CNN design tools. All these criteria play an important role in strategically defining trade-offs in order to direct the design flows towards the implementation of the appropriate goal. Among the different inference performances are: number of operations, time, accuracy (TOP-1 & TOP-5), hardware resources, latency and energy efficiency. The metrics used for inference performance evaluation of FPGA-based CNN models are defined below

1. Number of operations.

The number of operations is used to determine the complexity of a CNN model and the hardware load. The total number of operations (multiply-accumulate, divide, compare) required to compute the outputs for all layers of a CNN model can be calculated from the layer configurations applied (convolution kernel, number of layers, etc.) during the design phase.

2. FPGA hardware resources.

The number of hardware parameters (DSP / LUT / FF / BRAM) used in the inference architecture of FPGAs compared to the total hardware parameters available on the FPGA. The resource consumption is an indicator that allows to evaluate the optimization efficiency of the development tools on a target platform.

3. Energy consumption.

The power consumption will represent the power (W) required to perform image inference on an FPGA.

4. Energy efficiency;

Energy efficiency is described as a better measure of performance for a given energy cost. It is measured in "Operations / Second" per Watt (OPS / W), or Operations / Joule (OP / J), since $1W = 1J / s$. The energy cost is defined as the average number of operations that can be performed within an energy limit. It comes from two elements: computation and memory accesses.

5. Latency (ms)

Latency represents the time (ms) required to run a CNN model on an FPGA for a single image. The time will be measured from the start to the end of the inference propagation for all images in the test.

6. Throughput (frames / s)

Throughput is measured in GOPS (Giga Operations Per Second). Throughput represents the number of frames processed in one second by an FPGA. The throughput will be measured by the number of image batches executed during their latency time.

7. Number of operations per second (GOP / s)

A number of (Giga) operations per second that a type of hardware (FPGA, GPU) can perform is defined as the number of operations (MACC) that must be performed on the throughput.

8. Types of hardware architectures

FPGAs can offer flexible design and real-time performance. Many studies have been done to improve the optimization of these performances. Three types of accelerator architectures can be defined based on how the CNN layers are managed on the FPGA hardware. These three types of architectures are:

- Streaming accelerator architecture.
- Dedicated accelerator architecture.
- Programmable coprocessor architecture.

A. Streaming Accelerator:

For a streaming accelerator, all CNN layers are executed separately on an FPGA. In this type of accelerators, the results of each layer will be streamed to the input of the next layer. The different parameters are loaded once in a global memory and then they will be reused locally. Figure 4.3 shows the architecture of a streaming accelerator.

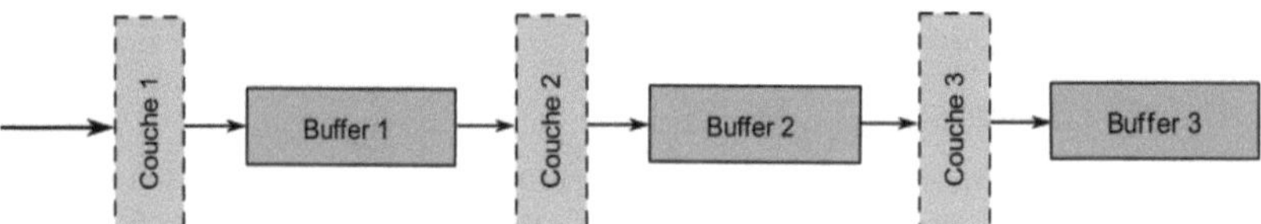

Figure 4.3- Streaming accelerator.

B. Dedicated accelerator

A dedicated accelerator (dedicate) extracts the layer classes on an FPGA and schedules the execution of these layers. A controller orchestrates the execution. It determines which layers to execute and when to load the parameters. The loading is usually done from global memory to local memory. Figure 4.4 shows the architecture of a dedicated accelerator.

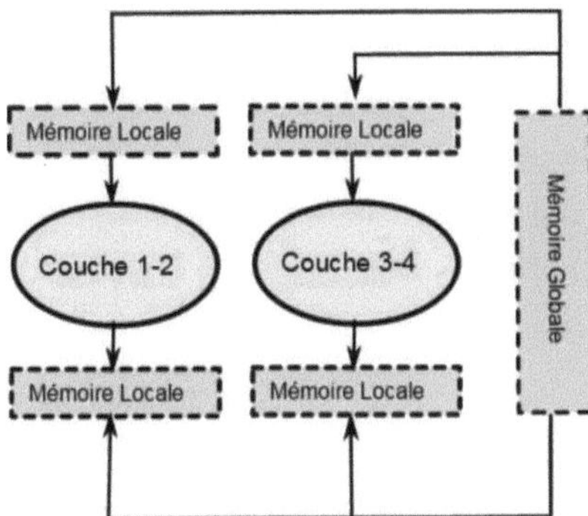

Figure 4.4- Dedicated accelerator.

C. Programmable coprocessor accelerator

In a programmable coprocessor architecture, MACC operations of different layers are executed systolic. A controller controls the execution of the layers. This controller manages the inputs and outputs of the systolic calculations. Some implementations use a processor as a controller or to execute intensive layers without doing the computation of fully connected or softmax layers. Figure 4.5 shows the architecture of a programmable coprocessor accelerator.

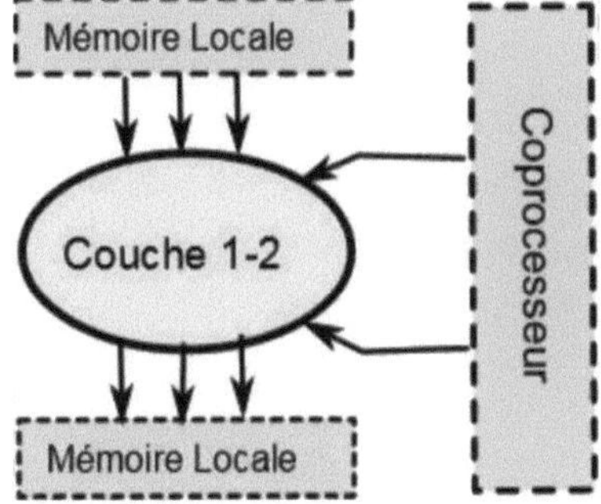

Figure 4.5 - programmable coprocessor.

9. Suggestions for future work :

Generally, it is preferable to do the training phase offline on GPU support in order to minimize the development and learning time. The use of development environments speeds up the implementation time. Software designs have focused primarily on compressing the model to reduce the number of weights or the size of bits used for each pixel or weight, resulting in reduced computational and storage complexity. Exploiting binary or fixed point coding allows for performance and resource optimization. Recent work has attempted to replace the neuron and weight representation with fixed point data instead of the commonly used floating point representation. The use of smaller bit size for pixels and weights contributes to the reduction of bandwidth, hardware cost and storage requirements of DNN systems.

Architectures and Accelerators of CNNs.

In this chapter, we will first cover and discuss CNNs based on FPGA implementations. We will review recent techniques using FPGAs for accelerating deep learning networks.

This in-depth study illustrates the use of various techniques such as exploiting parallelism, efficient use of internal memory to maximize data reuse, operation pipelining and efficient use of data sizes to minimize memory footprint and, to optimize the use of FPGA resources.

For each architecture and accelerator, we extract and describe the strengths and keys used to maximize performance and throughput in the acceleration process. The strengths determined through the study conducted can serve as a reference for researchers and designers in the field of CNNs.

I.   FPGA-based accelerators :

1. Optimization of FPGA-based accelerator design:

In [88], Chen. Zhang et al implemented a CNN accelerator on FPGA. They optimized the CNN computation and memory access. Under a working frequency of 100 MHz, this implementation achieves a maximum performance of 61.62 GFLOPS.

As shown in Figure 5.1, an FPGA-based CNN accelerator design [88] is composed of several main components, which are processing elements (PEs), an on-chip buffer, external memory, and an on/off-chip interconnect.

A PE is the basic computing unit for convolution. All data to be processed is stored in an external memory. Due to the limitation of on-chip resources, the data is first cached in on-chip buffers before being transmitted to the PEs.

Double buffers are used to cover the computation time with the data transfer time. The on-chip interconnect is dedicated to the data communication between the PEs and the on-chip buffer banks.

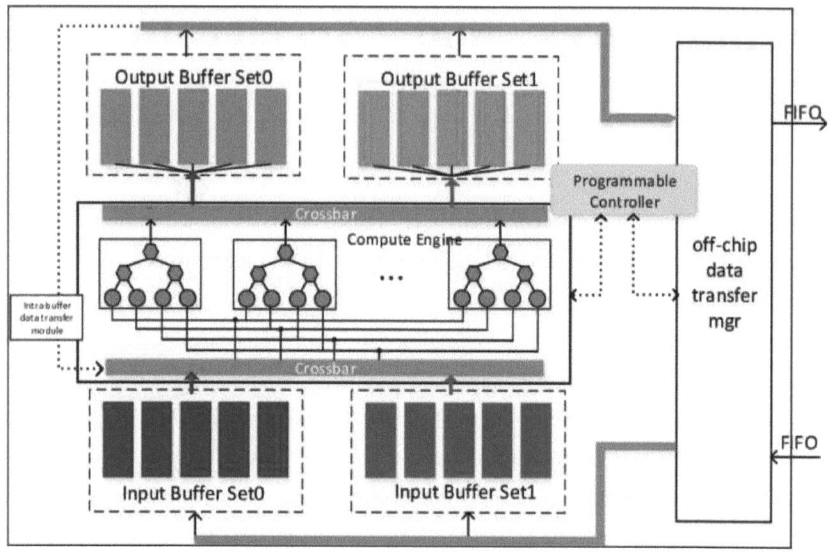

Figure 5.1 - Architecture of the Zhang et al. accelerator [88].

2. Caffeine: Scalable accelerator architecture on FPGA.

In [89], the authors implemented an efficient and reusable CNN / DNN FPGA accelerator that maximized FPGA computational capacity using a pipeline method similar to [88], authors C. Zhang et al used configurable finite state machines to schedule the execution of layer-specific instructions, which maximized memory bandwidth utilization by reordering memory accesses in the convolution. To realize a scalable convolution accelerator design, the authors designed a massive number of parallel PEs to improve the computational performance. They organized these PEs as a systolic array to alleviate the synchronization problems during synthesis of a large design. As a result, this hardware accelerator can achieve high performance of 96 GFLOPS for the convolution layer. As shown in Figure 5.2 , an overview of the scalable accelerator architecture, with corresponding buffers. Each PE is an arithmetic multiplication of the input feature map pixels and the corresponding weights. An array of adder trees summarizes the convolution results.

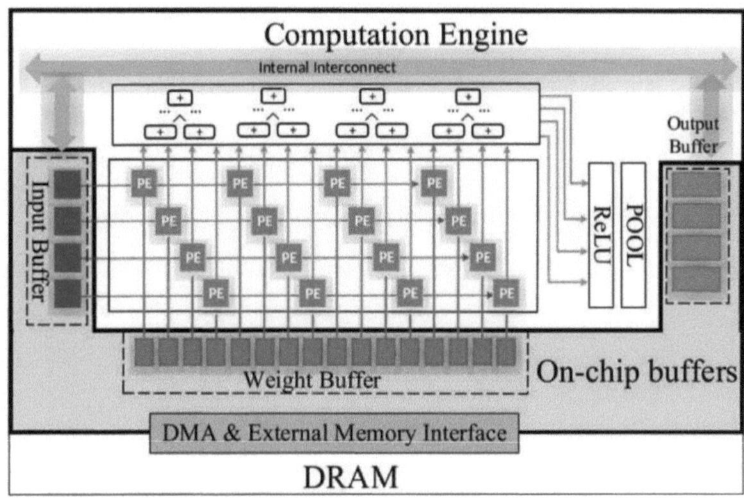

Figure 5.2 - "Caffeine" evolutionary accelerator architecture [89].

3. DLAU: a scalable deep learning accelerator on FPGA

In [91], Chao Wang et al proposed a hardware accelerator called "DLAU" for deep learning neural networks. The DLAU is described as a flexible and adaptable stand-alone unit to match with different applications. The suggested DLAU accelerator used three pipelined processing units to improve throughput. A main computational TMMU that reads the total weight data via DMA, performs the computations, and then transfers the results of the sum of the intermediate parts to the PSAU. The second "PSAU" collects the partial sums and performs the accumulation. When the accumulation is completed, the results will be transferred to AFAU. The third unit "AFAU" performs the activation function. Floating point coding was used for addition and multiplication operations. This DLAU could achieve up to 36.1x the acceleration of an Intel core2 processor for a 256×256 array size and consumes nearly 167 DSPs. Figure 5.3 shows the DLAU model.

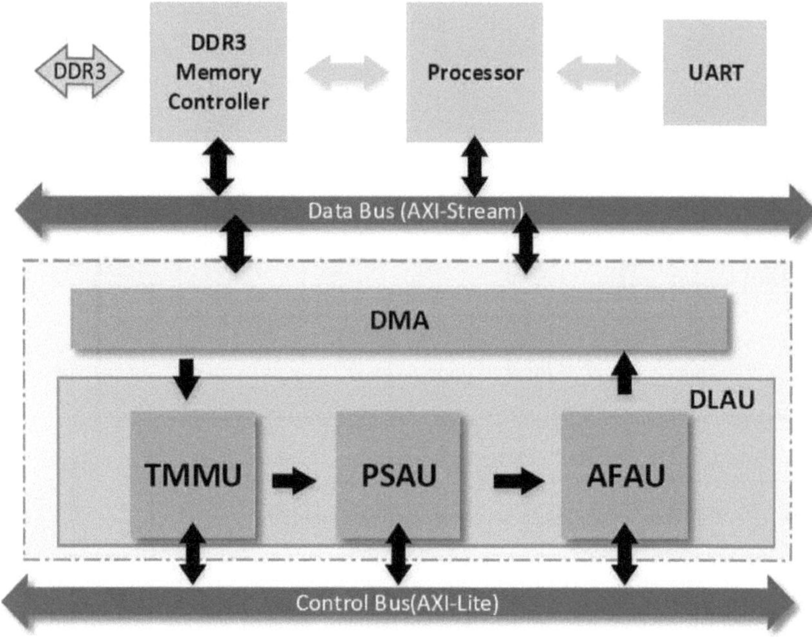

Figure 5.3- Architecture of the DLAU accelerator[91].

4. Deep CNN architecture using threshold neuron pruning on FPGA

In [92], the authors presented a technique for pruning neurons. In this work, the weight memory is realized by an on-chip memory embedded in FPGA with the aim of achieving fast memory access. In this paper, a fully connected layer (FC) circuit with a parallel input and a sequential output is proposed. Figure 5.4 shows the circuit architecture for a fully connected SIPO layer from [92].

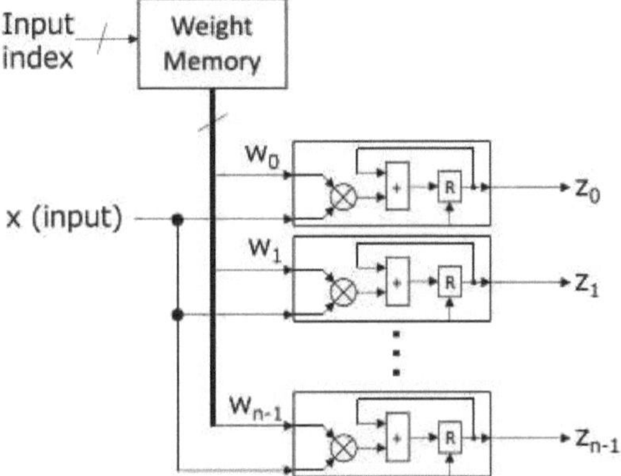

Figure 5.5 - Circuit architecture for a fully connected SIPO layer from [92].

Experimental results showed that the application of neural pruning, as for the fully connected layer on the VGG-11 CNN leads to the reduction of the number of neurons up to 89.3\% while maintaining 99\% accuracy. The fully connected layer implementation on the Digilent Inc. NetFPGA-1G-CML board showed that the FPGA support was 219.0x faster than the CPU (ARM Cortex A15 processor) and 12.5x faster than the GPU (Jetson TK1 Kepler). In addition, the performance per energy efficiency was 125.28x faster than the CPU and 17.88x faster than the GPUs. Depth-separable CNNs can significantly reduce the number of model parameters and improve computational speed, so it is naturally suitable for computing applications in mobile devices.

5. CNP: An FPGA-based processor for CNNs.

Farabet et al [93] presented an FPGA implementation of CNN called convolutional network processor (CNP) which uses a hardware convolver and a processor for data processing and control. The proposed architecture consists of the Virtex4 SX35 FPGA platform and an external memory. CNP has been implemented for a face detection system with LeNet-5 architecture and power consumption less than 15 watts.

In the proposed architecture, the authors have used a 2D kernel convolution module of size 3 x 3, output buffers (FIFO) between FPGA and external memory in both directions to ensure stable data flow. The proposed implementation exploits CONV layer parallelism to accelerate the

CNN computational engine while using a large number of DSPs and MACCs on FPGA. The implementation of this 2D convolution module is shown in Figure 5.6.

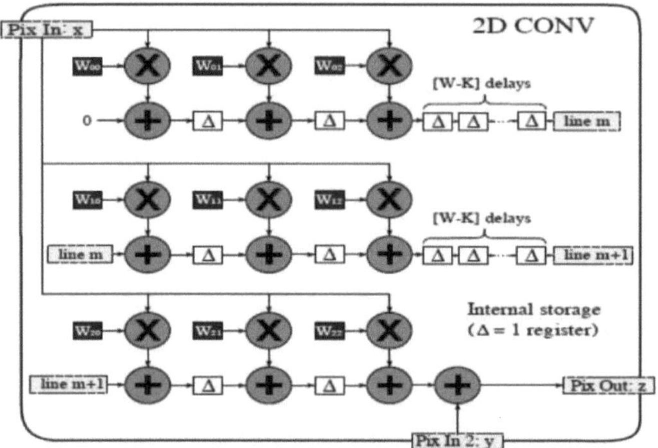

Figure 5.6 - A 2D Convolution for K = 3 [93].

where x is the input feature map data, K = kernel width = kernel height, W is the width of the input image, Wij is the value of the weight in the convolution kernel, y is the partial result to be combined with the current result and z is the result to the output. It can be seen that the proposed convolutional module performs K2 MACCs operations simultaneously in each clock cycle.

6. Design of an efficient accelerator of depth-separable CNNs on FPGA

In [94], a depth-separable convolutional neural network accelerator is proposed. In this accelerator, the layers are executed simultaneously in a pipelined manner to improve the throughput and performance of the system. To exploit the parallelism in CNNs, several hardware optimization strategies are adopted in the convolutional layer. The proposed accelerator achieves a good balance between speed, resource utilization and power consumption. For 2D convolution, the multiplication has been performed using parallel multipliers. The addition of all the results of the parallel multipliers is performed using an adder tree. It is noted that the MACC operations were 16 bits in fixed point coding. In [94], the intermediate data of each layer is buffered into the on-chip memory to reduce the required memory bandwidth and data access power consumption.

The proposed accelerator has been implemented and evaluated on Intel Arria 10 with Intel Quartus Prime 16.0 for synthesis and ModelSim for simulation. The results show that the proposed accelerator achieves a performance of 98.9 GOP/s, up to 17.6× faster and 29.4× lower power than CPU and GPU based implementations respectively. Figure 5.7 represents An overview of the accelerator architecture of [94] and its different units and modules.

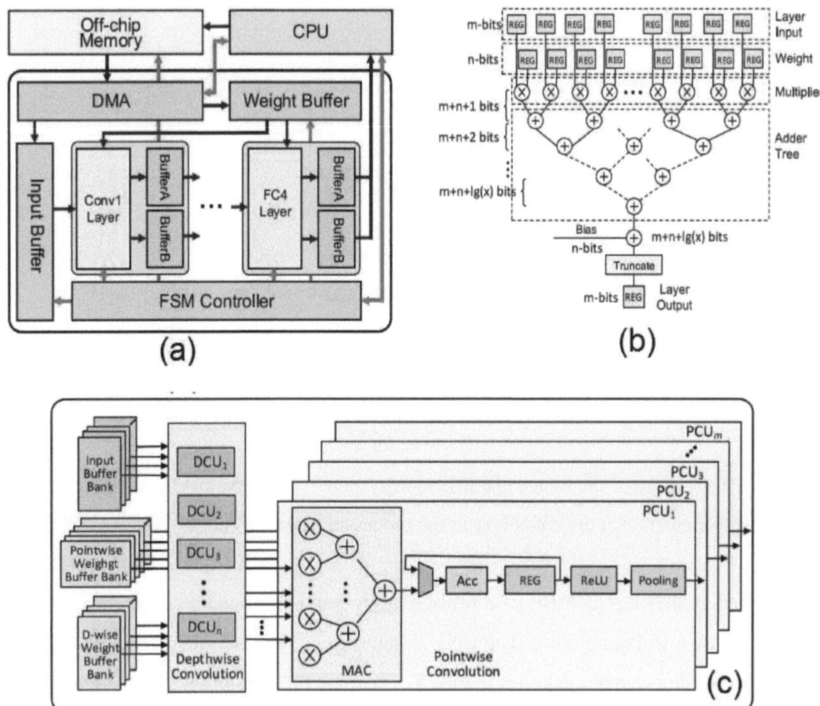

Figure 5.7 - a) An overview of the accelerator architecture, (b) MAC unit, (c) Depth-separable convolutional layer computation engine [94].

7. The integrated FPGA platform for the convolutional neural network :

In [71], Qiu et al. proposed an FPGA design to accelerate CNNs for a large-scale image classification application on embedded systems. The authors focused on accelerating the

CONV and FC layers, since they are considered the most computationally and memory-centric operations in CNNs. The proposed accelerator reduces the resource consumption by using a specific design of the convolver hardware module. In addition, the authors have applied data quantization to reduce the memory and bandwidth of CNN.

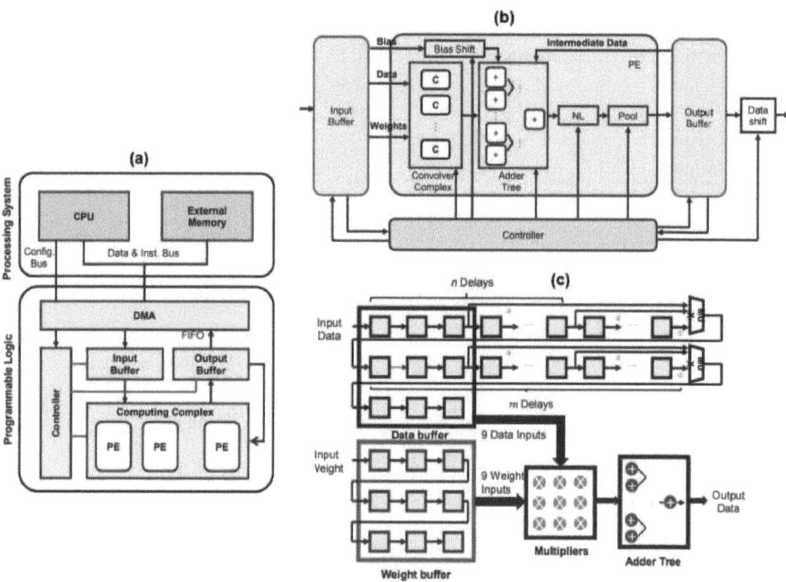

Figure 5.8 - The design of the image classification system [71]: (a) the overall architecture; (b) the processing element; (c) the convolver in the processing element.

The proposed architecture consists of a processing system (CPU) and a programmable logic (FPGA) as shown in Figure 5.8.a. The CNN computations are performed through a special design of processing element modules in FPGA. The main modules of the processing element are convolver complex, maximal pooling, nonlinearity, data shifting and an adder tree, as shown in Figure 5.8.b. The convolver complex is shown in Figure 5.8.c, to perform convolution operations as well as to compute the multiplication of the FC layer. The implemented pooling layer incorporates a window of size 2. The activation function is applied using the nonlinearity module. The adder tree accumulates the partial sums generated by the convolvers. Finally, the shift modules are responsible for the dynamic quantization.

Most of the works such as [95] [90] already done for the acceleration of CNNs on FPGAs have been mainly focused on optimizing the hardware resources in order to optimize the convolution layer since it is computationally related. Other works [86] [87] have been mainly focused on the optimization of external memory transfers. Acceleration or optimization of a single layer such as the convolution layer can lead to limitations in the system where the other CNN layers remain unaccelerated, which prevents the FPGA support from achieving high energy efficiency. The unaccelerated layer that is fully connected can become the bottleneck of the system. The need for a large amount of data causes the creation of data movement from off-chip memory which means an increase in the number of external memory accesses.

Data movement involves higher power consumption than computation. Most of the performance gains in FPGA resources such as acceleration, throughput and power efficiency are achieved with optimized, flexible and configurable architecture, data and computation binarization, off-chip training, pipeline parallelism and preconfigured topology.

II. Analysis of different CNN accelerators:

In Table 4.1, we summarize the FPGA-based deep learning network acceleration techniques. For each technique, we list the year of publication, the strengths used for acceleration, the deep learning model accelerated, the number of operations required per image, the FPGA platform used for implementation, the accuracy used, the frequency, the type of LUT for the platform used, the performance in GOPS, the acceleration with respect to a given baseline model, and finally the energy efficiency (GOPS / W).

Analysis of different CNN accelerators can provide insight into design choices that can lead to efficient and optimized neural networks for a real-time application. In the context of FPGA implementation, the preferred performance parameters are typically acceleration, throughput, FPGA resources, power consumption and energy efficiency. After the analysis, we clarified and simplified the representation of the implementation parameter by grouping the interesting solutions according to the choice of strategies and development flow, as well as the performance parameters for the optimization and implementation of FPGA-based deep learning networks. The study carried out shows that most of the works have adopted a preconfigured topology, off-chip training, optimized phase and a configurable and flexible architecture.

The development flow describes the importance of using the framework with a high-level language as well as applying pipeline parallelism. Indeed, most of the work has advocated deep learning with Xilinx hardware development tools like Vivado and Vivado HLS. Most of the performance gains in FPGA resources, acceleration, throughput and power efficiency are achieved with pipeline parallelism, optimized, flexible and configurable architecture, off-chip training and preconfigured topology. Binary coding is adopted for FPGA resources and speedup gains, while in energy efficiency and throughput gains, 16-bit fixed-point coding is frequently used. Therefore, the best way to implement deep learning on FPGAs is to perform the training phase on GPU support, not only to minimize the development time, but also to overcome the large size that the most efficient network can reach. FPGA offers excellent support for inference, as it allows for power savings due to low working frequency. In fact, available tools like Vivado HLS simplify implementation on FPGA by supporting a high-level language like C or OpenCL (Vivado SDAccel) and synthesizable data types. The use of frameworks speeds up the implementation time while the integration framework process in the optimization phase is desirable. Arithmetic can be provided with fixed point, and also with binary coding, which allows for performance and resource optimization.

Table 4.1 - FPGA accelerator performance summary

Architecture	Ref.	Models	GOP	Platform	Prec.	Freq (MHz)	GOPS	Acceleration	Baseline	$\frac{GOPS}{W}$	Highlights
DLAU	[91]	DNN	N/A	Zynq XC7Z020	FP 48	200	N/A	36.1x	2.3 GHz Intel Core2	N/A	• Pipelined treatment units • The tiles, • FIFO stamps.
Embedded FPGA	[71]	VGG16	3.76	Zynq XC7Z045	INT 16	150	136.97	1.4x	2.9 GHz Intel Xeon	14.22	• Quantification of data. • Arrangement of data.
Roofline-based FPGA	[88]	AlexNet	1.33	Virtex7 VX485T	FP 32	100	61.62	17.42x	2.2 GHz Intel Xeon	3.31	• Systolic network architecture, • Loop unwinding, • Double buffer, • Pipeline,
Caffeine	[89]	AlexNet	1.46	Virtex7 VX690T	INT 16	150	354	9.7x	Intel E5-2609 CPU, 1.9 GHz	13.62	• Systolic network architecture, • Loop unwinding, • Double buffer, • Pipeline,
							266	7.3x		10.64	
		VGG-16	30.9	Xilinx KU060	INT 8	200	1,171.7	29x		45.07	
							165	4.2x		N/A	

65

Conclusion

In this book, we have introduced the general basics of deep learning and classification, prerequisites have been carefully chosen to allow a quick and easy understanding of the basic concepts and contexts needed for deep learning and CNNs training. In addition, this book has covered recent developments in CNNs and deep learning models as well as techniques exploited for performance improvement. Their use has enabled several authors to achieve good classification performance. This work discusses innovative hardware-based techniques and solutions for integrating deep learning into energy-constrained embedded devices. In addition, we have presented recent and interesting developments in the field of DNN acceleration and, in particular, CNNs on FPGAs. The highlights determined through the study carried out in this work can serve as a reference for researchers and designers in the field of CNNs. In addition, we have provided, in this book, a brief overview of acceleration techniques for deep learning algorithms highlighting their importance, key operations and applications. We are interested, in particular, in CNNs since they have various architectures and accelerators and broad applications in the field of image recognition. The main body of this chapter has been a review of the techniques used in accelerating CNNs on FPGAs.

This book, also introduced the use of tools to generate high level circuit representations and RTL (Register Transfer Level) scripts that not only help to automate the design process, but also explore the design space and suggest efficient hardware. Overall, this book explores and analyzes the main features used by various FPGA-based CNN acceleration techniques and delivers recommendations for improving the efficiency of using FPGAs in CNN acceleration.

Bibliography

- *[1] F. Rosenblatt (1958), The perceptron: a probabilistic model for information storage and organization in the brain,*

- [2] WS McCulloch, W Pitts - A logical calculus of the ideas immanent in nervous activity The bulletin of mathematical biophysics, 1943 - Springer.

- [3] Rumelhart, D., Hinton, G. & Williams, R. Learning representations by back-propagating errors. *Nature* **323,** 533-536 (1986). https://doi.org/10.1038/323533a0

- [4] Ruder, Sebastian. "An overview of gradient descent optimization algorithms." ArXiv abs/1609.04747 (2016): n. pag.

- [5] Richard S. Sutton. Two problems with backpropagation and other steepest-descent learning procedures for networks, 1986

- [6] John Duchi, Elad Hazan, and Yoram Singer. Adaptive Subgradient Methods for Online Learning and Stochastic Optimization. Journal of Machine Learning Research, 12:2121-2159, 2011.

- [7] Diederik P. Kingma and Jimmy Lei Ba. Adam: a Method for Stochastic Optimization. International Conference on Learning Representations, pages 1-13, 2015.

- [8] Matthew D. Zeiler. ADADELTA: An Adaptive Learning Rate Method. arXiv preprint arXiv:1212.5701, 2012.

- [9] Ning Qian. On the momentum term in gradient descent learning algorithms. Neural networks : the official journal of the International Neural Network Society, 12(1):145-151, 1999.

- [10] Jalled, Fares. "Face Recognition Machine Vision System Using Eigenfaces." ArXiv abs/1705.02782 (2017): n. pag.

- [11] R. E. Kalman. A new approach to linear filtering and prediction problems. ASME Journal of Basic Engineering, 1960, 82(1): 35-45, https ://doi.org/10.1115/1.3662552, Published Online : March 1, 1960

- [12] P. Viola and M. Jones. Rapid object detection using a boosted cascade of simple features. In Proceedings of the 2001 IEEE Computer Society Conference on Computer Vision and Pattern Recognition. CVPR 2001, volume 1, pages I-I, Dec 2001.
- [13] Goodfellow IJ, Warde-Farley D, Mirza M, Courville AC, Bengio Y (2013) Maxout networks. In: Proceedings of the 30th international conference on machine learning (ICML 2013), volume 28 of JMLR proceedings, pp 1319-1327. http ://jmlr.org/
- [14] Clevert, Djork-ArnÃ© Unterthiner, Thomas Hochreiter, Sepp. (2015). Fast and Accurate Deep Network Learning by Exponential Linear Units (ELUs). Under Review of ICLR2016 (1997).
- [15] Ioffe and C. Szegedy, Batch normalization: Accelerating deep network training by reducing internal covariate shift. In ICML, 2015.
- [16] Srivastava Nitish, Hinton Geoffrey, Krizhevsky Alex, Sutskever Ilya, Salakhutdinov Ruslan (2014) Dropout: a simple way to prevent neural networks from overfitting. J Mach Learn Res 15 :1929-1958.
- [17] Glorot, X., & Bengio, Y. (2010, March). Understanding the difficulty of training deep feedforward neural networks. In Proceedings of the thirteenth international conference on artificial intelligence and statistics (pp. 249-256).
- [18] He, Kaiming, et al. "Delving deep into rectifiers: Surpassing human-level performance on imagenet classification." Proceedings of the IEEE International Conference on Computer Vision. 2015.
- [19] Ananthram, Aditya. 'Random Initialization For Neural Networks: A Thing Of The Past. Towards Data Science'. 25 Feb 2018.Available: towardsdatascience.com/random-initialization-for-neural-networks-a-thing-of-the-past-bfcdd806bf9e
- [20] Hochreiter S., Schmidhuber J., "Long Short-Term Memory", Neural Comput. 9, 8, 1735-1780, November, 1997.
- [21]   G. E. Hinton, N. Srivastava, A. Krizhevsky, I. Sutskever, and R. Salakhutdinov. Improving neural networks by preventing co-adaptation of feature detectors. CoRR, abs/1207.0580, 2012.

- [22]   D. Ciresan, U. Meier, J. Schmidhuber, "Multi-column deep neural networks for image classification", CoRR, vol. abs/1202.2745, 2012, [online] Available: http://arxiv.org/abs/1202.2745.

- [23] K. Gregor, Y. LeCun, "Emergence of complex-like cells in a termporal product network with local receptive fields", CoRR, vol. abs/1006.0448, 2010, [online] Available: http://arxiv.org/abs/1006.0448.

- [24]   K. Jarrett, K. Kavukcuoglu, M. A. Ranzato, Y. LeCun, "What is the best multi-stage architecture for object recognition? ", Proc. IEEE Int. Conf. Comput. Vis. pp. 2146-2153, Sep. 2009.

- [25] Y. LeCun, K. Kavukcuoglu, C. Farabet, "Convolutional networks and applications in vision," Proc. IEEE Int. Symp. Circuits Syst. pp. 253-256, Jun. 2010.

- [26]   M. D. Zeiler, R. Fergus, "Stochastic pooling for regularization of deep convolutional neural networks", CoRR, vol. abs/1301.3557, 2013,\ [online] Available: http://arxiv.org/abs/1301.3557.

- [27]   K. He, X. Zhang, S. Ren, J. Sun, "Spatial pyramid pooling in deep convolutional networks for visual recognition," Proc. Eur. Conf. Comput. Vis. pp. 346-361, 2014.\

- [28]   T. Chan, K. Jia, S. Gao, J. Lu, Z. Zeng, Y. Ma, "PCANet: A simple deep learning baseline for image classification? ", CoRR, vol. abs/1404.3606, 2014, [online] Available:_http://arxiv.org/abs/1404.3606.

- [29]   C. Lee, P. Gallagher, Z. Tu, "Generalizing pooling functions in convolutional neural networks: Mixed gated and tree", CoRR, vol. abs/1509.08985, 2015, [online] Available: https://arxiv.org/abs/1509.08985.

- [30]   J. Springenberg, A. Dosovitskiy, T. T. Brox, M. Riedmiller, "Striving for simplicity: The all convolutional net", CoRR, vol. abs/1412.6806, 2014, [online] Available: http://arxiv.org/abs/1412.6806.

- [31]   K. Gregor, Y. LeCun, "Emergence of complex-like cells in a termporal product network with local receptive fields", CoRR, vol. abs/1006.0448, 2010, [online] Available: http://arxiv.org/abs/1006.0448.

- [32]   D. Yoo, S. Park, J. Lee, I. Kweon, "Multi-scale pyramid pooling for deep convolutional representation", Proc. IEEE Workshop Comput. Vis. Pattern Recognit. pp. 1-5, Sep. 2015.

- [33]   B. Graham, "Fractional max-pooling", CoRR, vol. abs/1412.6071, 2014, [online] Available: https://arxiv.org/abs/1412.60712.

- [34]. Lecun, Y.; Bottou, L.; Bengio, Y.; Haffner, P. (1998). "Gradient-based learning applied to document recognition" (PDF). Proceedings of the IEEE. 86 (11): 2278-2324. doi:10.1109/5.726791.

- [35] Krizhevsky, Alex & Sutskever, Ilya & Hinton, Geoffrey. (2012). ImageNet Classification with Deep Convolutional Neural Networks. Neural Information Processing Systems. 25. 10.1145/3065386.

- [36] Zeiler, Matthew & Fergus, Rob. (2013). Visualizing and Understanding Convolutional Neural Networks. ECCV 2014, Part I, LNCS 8689. 8689. 10.1007/978-3-319-10590-1_53.

- [37] Simonyan, Karen & Zisserman, Andrew. (2014). Very Deep Convolutional Networks for Large-Scale Image Recognition. arXiv 1409.1556.

- [38] C. Szegedy et al, "Going deeper with convolutions," 2015 IEEE Conference on Computer Vision and Pattern Recognition (CVPR), Boston, MA USA, 2015, pp. 1-9. doi:10.1109/CVPR.2015.7298594.

- [39] He X. Zhang, S. R. and Sun, J. Deep residual learning for image recognition", ArXiv preprint arXiv:1512.03385, 2015.

- [40] S. Zagoruyko and N. Komodakis. Wide residual networks. In BMVC, 2016.\

- [41] He, Kaiming & Zhang, Xiangyu & Ren, Shaoqing & Sun, Jian. (2016). Identity Mappings in Deep Residual Networks. 9908. 630-645. 10.1007/978-3-319-46493-0_38.

- [42] Saining Xie, Ross Girshick, Piotr Dollár, Zhuowen Tu, Kaiming He , Aggregated Residual Transformations for Deep Neural Networks.2016.

- [44] Gao Huang, Zhuang Liu, Laurens van der Maaten, Kilian Q. Weinberger, Densely Connected Convolutional Networks ,2017.

- [44] Min Lin, Qiang Chen, Shuicheng Yan, Network In Network, ICLR 2014.

- [45] Alaeddine, H., Jihene, M. Deep network in network. *Neural Comput & Applic* **33,** 1453-1465 (2021). https://doi.org/10.1007/s00521-020-05008-0.

- [46] J. Hu, L. Shen and G. Sun, "Squeeze-and-Excitation Networks," *2018 IEEE/CVF Conference on Computer Vision and Pattern Recognition*, 2018, pp. 7132-7141, doi: 10.1109/CVPR.2018.00745.

- [47] Iandola, Forrest & Han, Song & Moskewicz, Matthew & Ashraf, Khalid & Dally, William & Keutzer, Kurt. (2016). SqueezeNet: AlexNet-level accuracy with 50x fewer parameters and <0.5MB model size.

- [48] Larsson, G., Maire, M., Shakhnarovich, G. (2016). Fractalnet: Ultra-deep neural networks without residuals. arXiv

- preprint arXiv :1605.07648.\

- [49] Howard, Andrew G., Menglong Zhu, Bo Chen, Dmitry Kalenichenko, Weijun Wang, Tobias Weyand, Marco Andreetto, and Hartwig Adam. "Mobilenets: Efficient convolutional neural networks for mobile vision applications." arXiv preprint arXiv :1704.04861 (2017).\

- [50] Zhang, X., Zhou, X., Lin, M., Sun, J. (2018). Shufflenet: An extremely efficient convolutional neural network for mobile devices. In Proceedings of the IEEE conference on computer vision and patternrecognition (pp. 6848-6856).

- [51] G. Lacey, G. W. Taylor, and S. Areibi, "Deep learning on FPGAs: Past, present, and future," arXiv preprint arXiv :1602.04283, 2016. 6.

- [52] A. Munshi, "The OpenCL specification," 2009 IEEE Hot Chips 21 Symposium (HCS), 2009, pp. 1-314, doi: 10.1109/HOTCHIPS.2009.7478342.

- [53] J. E. Stone, D. Gohara, and G. Shi, "OpenCL: A parallel programming standard for heterogeneous computing systems," Computing in science engineering, vol. 12, no. 3, pp. 66-73, 2010. 6.

- [54] Omondi, Amos Rajapakse, Jagath. (2006). FPGA Implementations of Neural Networks. 10.1007/0-387-28487-7.

- [55] O. Nomura and T. Morie, "Projection-field-type VLSI convolutional neural networks using merged/mixed analog-digital approach," in International Conference on Neural Information Processing. Springer, 2007, pp. 1081-1090. 2 .

- [56] M. Sankaradas, V. Jakkula, S. Cadambi, S. Chakradhar, I. Durdanovic, E. Cosatto, and H. P. Graf. A massively parallel coprocessor for convolutional neural networks. In Application-specific Systems, Architectures and Processors, 2009. ASAP 365 2009. 20th IEEE International Conference on, pages 53âAS60. IEEE, 2009.

- [57] T. M. Chilimbi, Y. Suzue, J. Apacible, and K. Kalyanaraman, "Project Adam: Building an efficient and scalable deep learning system." in OSDI, vol. 14, 2014, pp. 571-582. 2 .

- [58] Y. LeCun, B. Boser, J. S. Denker, D. Henderson, R. E. Howard, W. Hubbard, and L. D. Jackel, "Backpropagation applied to handwritten zip code recognition," Neural computation, vol. 1, no. 4, pp. 541-551, 1989. 3.

- [59] A. Yazdanbakhsh, J. Park, H. Sharma, P. Lotfi-Kamran, and H. Esmaeilzadeh, "Neural acceleration for gpu throughput processors," in Proceedings of the 48th International Symposium on Microarchitecture. ACM, 2015, pp. 482-493. 3.

- [60] G. Hinton, L. Deng, D. Yu, G. E. Dahl, A.-r. Mohamed, N. Jaitly, A. Senior, V. Vanhoucke, P. Nguyen, T. N. Sainath, et al. "Deep neural networks for acoustic modeling in speech recognition: The shared views of four research groups," IEEE Signal processing magazine, vol. 29, no. 6, pp. 82-97, 2012. 3.

- [61] Y. Jia, E. Shelhamer, J. Donahue, S. Karayev, J. Long, R. Girshick, S. Guadarrama, and T. Darrell, "Caffe: Convolutional architecture for fast feature embedding," in Proceedings of the 22nd ACM international conference on Multimedia. ACM, 2014, pp. 675-678. 3, 14, 25.

- [62] A. Vasudevan, A. Anderson, and D. Gregg, "Parallel multi channel convolution using general matrix multiplication," in Application-specific Systems, Architectures and Processors (ASAP), 2017 IEEE 28th International Conference on. IEEE, 2017, pp. 19-24. 3.

- [63] Janardan Misra, Indranil Saha, Artificial neural networks in hardware: A survey of two decades of progress, Neurocomputing, Volume 74, Issues 1â€"3, 2010, Pages 239-255, ISSN 0925-2312, https ://doi.org/10.1016/j.neucom.2010.03.021. 3.

- [64] H. Esmaeilzadeh, A. Sampson, L. Ceze, and D. Burger, "Neural acceleration for general-purpose approximate programs," in Proceedings of the 2012 45th Annual IEEE/ACM International Symposium on Microarchitecture. IEEE Computer Society, 2012, pp. 449-460. 3.

- [65] S. Han, X. Liu, H. Mao, J. Pu, A. Pedram, M. A. Horowitz, and W. J. Dally, "Eie: efficient inference engine on compressed deep neural network," in Computer Architecture (ISCA), 2016 ACM/IEEE 43rd Annual International Symposium on. IEEE, 2016, pp. 243-254. 3.
- [66] L. Du, Y. Du, Y. Li, J. Su, Y.-C. Kuan, C.-C. Liu, and M.-C. F. Chang, "A reconfigurable streaming deep convolutional neural network accelerator for internet of things," IEEE Transactions on Circuits and Systems I: Regular Papers, vol. 65, no. 1, pp. 198-208, 2018. 3.
- [67] Vanderbauwhede, Wim Benkrid, K.. (2013). High-performance computing using FPGAs. 10.1007/978-1-4614-1791-0.
- [68] A. Putnam, A. M. Caulfield, E. S. Chung, D. Chiou, K. Constantinides, J. Demme, H. Esmaeilzadeh, J. Fowers, G. P. Gopal, J. Gray, et al. "A reconfigurable fabric for

accelerating large-scale datacenter services," ACM SIGARCH Computer Architecture News, vol. 42, no. 3, pp. 13- 24, 2014. 3, 13.

- [69] V. Sze, Y.-H. Chen, J. Emer, A. Suleiman, and Z. Zhang, "Hardware for machine learning: challenges and opportunities," in Custom Integrated Circuits Conference (CICC), 2018 IEEE. IEEE, 2018, pp. 1-8. 6.

- [70] N. Suda, V. Chandra, G. Dasika, A. Mohanty, Y. Ma, S. Vrudhula, J.-s. Seo, and Y. Cao, "Throughput-optimized opencl-based FPGA accelerator for large-scale convolutional neural networks," in Proceedings of the 2016 ACM/SIGDA International Symposium on Field-Programmable Gate Arrays. ACM, 2016, pp. 16-25. 4, 5, 6, 14, 16, 17, 20, 21, 22, 29, 30, 31, 32, 34.

- [71] J. Qiu et al, "Going deeper with embedded FPGA platform for convolutional neural network," in Proc. ACM/SIGDA Int. Symp. Field- Programm. Gate Arrays (FPGA), New York, NY, USA, 2016, pp. 26-35.

- [72] S. Han, J. Kang, H. Mao, Y. Hu, X. Li, Y. Li, D. Xie, H. Luo, S. Yao, Y. Wang, et al, "Ese: Efficient speech recognition engine with sparse lstm on FPGA," in Proceedings of the 2017 ACM/SIGDA International Symposium on Field-Programmable Gate Arrays. ACM, 2017, pp. 75- 84. 6.

- [73] Yangqing Jia, Evan Shelhamer, Jeff Donahue, Sergey Karayev, Jonathan Long, Ross Girshick, Sergio Guadarrama, and Trevor Darrell. Caffe: Convolutional architecture for fast feature embedding. arXiv preprint arXiv :1408.5093, 2014. 8.

- [74] Theano Development Team. Theano: A Python framework for fast computation of mathematical expressions. arXiv e-prints, abs/1605.02688, May 2016. 8.

- [75] Mart'ın Abadi, Ashish Agarwal, Paul Barham, Eugene Brevdo, Zhifeng Chen, Craig Citro, Greg S. Corrado, Andy Davis, Jeffrey Dean, Matthieu Devin, Sanjay Ghemawat, Ian Goodfellow, and et al. TensorFlow: Large-scale machine learning on heterogeneous systems, 2015. Software available from tensorflow.org. 8

- [76] Ross Girshick, Je. Donahue, Trevor Darrell, and Jitendra Malik. 2014. Rich feature hierarchies for accurate object detection and semantic segmentation. In Proceedings of the IEEE conference on computer vision and pa.ern recognition. 580–587.

- [77] Raghid Morcel, Haitham Akkary, Hazem Hajj, Mazen Saghir, Anil Keshavamurthy, Rahul Khanna, and Hassan Artail. 2017. Minimalist Design for Accelerating Convolutional Neural Networks for Low-End FPGA Platforms. In Field-Programmable Custom Computing Machines (FCCM), 2017 IEEE 25th Annual International Symposium on. IEEE, 196-196.

- [78] Roberto DiCecco, G. Lacey, J. Vasiljevic, P. Chow, G. Taylor, and S. Areibi. Caffeinated FPGAs: FPGA framework for convolutional neural networks. In 2016 International Conference on Field-Programmable Technology (FPT), pages 265-268, Dec 2016. 9, 14.

- [79] Andrew G Howard, Menglong Zhu, Bo Chen, Dmitry Kalenichenko, Weijun Wang, Tobias Weyand, Marco Andree.o, and Hartwig Adam. 2017. MobileNets: Ecient Convolutional Neural Networks for Mobile Vision Applications. (2017).

- [80] Yongming Shen, Michael Ferdman, and Peter Milder. 2017. Escher: A CNN Accelerator with Flexible Bu.ering to Minimize O.-Chip Transfer. In Proceedings of the 25th IEEE International Symposium on Field-Programmable Custom Computing Machines (FCCM€17). IEEE Computer Society, Los Alamitos, CA, USA.

- [81] Naveen Suda, Vikas Chandra, Ganesh Dasika, Abinash Mohanty, Yufei Ma, Sarma Vrudhula, Jae-sun Seo, and Yu Cao. 2016. . roughput-optimized OpenCL-based FPGA accelerator for large-scale convolutional neural networks. In Proceedings of the 2016 ACM/SIGDA International Symposium on Field-Programmable Gate Arrays. ACM, 16-25.

- [82] Yijin Guan, Hao Liang, Ningyi Xu, Wenqiang Wang, Shaoshuai Shi, Xi Chen, Guangyu Sun, Wei Zhang, and Jason Cong. 2017. FP-DNN: An Automated Framework for Mapping Deep Neural Networks onto FPGAs with RTL-HLS Hybrid Templates. In Field Programmable Custom Computing Machines (FCCM), 2017 IEEE 25th Annual International Symposium on. IEEE, 152-159.

- [83] Mr. Horowitz. [n. d.]. Energy table for 45nm process, Stanford VLSI wiki.[Online]. h.ps ://sites.google.com/site/seecproject.

- [84] Yijin Guan, Zhihang Yuan, Guangyu Sun, and Jason Cong. 2017. FPGA-based accelerator for long short-term memory recurrent neural networks. In Design Automation Conference (ASP-DAC), 2017 22nd Asia and South Pacic. IEEE, 629-634.

- [85] Yangqing Jia, Evan Shelhamer, Je Donahue, Sergey Karayev, Jonathan Long, Ross Girshick, Sergio Guadarrama, and Trevor Darrell. 2014. Caffe: Convolutional Architecture for Fast Feature Embedding. arXiv preprint arXiv:1408.5093 (2014).

- [86] S. Han, H. Mao, and W.J. Dally, "Deep compression: compressing deep neural network with pruning, trained quantization and huffman coding," 4th International Conference on Learning Representations, 2016.

- [87] M. Peemen, A. A. A. Setio, B. Mesman and H. Corporaal, "Memory-centric accelerator design for Convolutional Neural Networks," 2013 IEEE 31st International Conference on Computer Design (ICCD), Asheville, NC, 2013, pp. 13-19. doi: 10.1109/ICCD.2013.6657019.

- [88] Chen Zhang, Peng Li, Guangyu Sun, Yijin Guan, Bingjun Xiao, and Jason Cong. Optimizing FPGA-based accelerator design for deep convolutional neural networks. In Proceedings of the 2015 ACM/SIGDA International Symposium on Field Programmable Gate Arrays, FPGA '15, pages 161-170, New York, NY, USA, 2015. ACM. 9, 14.

- [89] Zhang, C., Fang, Z., Zhou, P., Pan, P., & Cong, J. (2016). Caffeine. Proceedings of the 35th International Conference on Computer-Aided Design - ICCAD '16. doi:10.1145/2966986.2967011.

- [90] Xuelei Li, Liangkui Ding, Li Wang, Fang Cao: FPGA accelerates deep residual learning for image recognition, technology, networking, electronic and automation control conference (itnec), 2017 ieee 2nd information.

- [91] C. Wang, Q. Yu, L. Gong, X. Li, Y. Xie, X. Zhou, DLAU: a scalable deep learning accelerator unit on FPGA, in: Proceedings of the IEEE Transactions on ComputerAided Design of Integrated Circuits and Systems, Volume 36, Issue 3, March 2017 March, doi: 10.1109/TCAD.2016.2587683 .

-
- [92] T. Fujii, S. Sato, H. Nakahara, M. Motomura, An FPGA realization of a deep convolutional neural network using a threshold neuron pruning, in: Proceed- ings of the ARC, in: LNCS, 10216, 2017, pp. 268-280, doi: 10.1007/978-3-319- 56258-223.

-
- [93] Farabet, Clement Poulet, Cyril Han, Jefferson Lecun, Yann. (2009). CNP : An FPGA-based processor for Convolutional Networks. FPL 09 : 19th International Conference on Field Programmable Logic and Applications. 10.1109/FPL.2009.5272559.

- [94] W. Ding, Z. Huang, Z. Huang, L. Tian, H. Wang, S. Feng, Designing efficient accelerator of depthwise separable convolutional neural network on FPGA, J. Syst. Archit. (2018) available online 28 December, doi: 10.1016/j.sysarc.2018.12.008 .

- [95] C. Farabet, B. Martini, P. Akselrod, S. Talay, Y. LeCun, E. Cu- lurciello, Hardware accelerated convolutional neural networks for synthetic vision systems, in : Proceedings of the IEEE International Symposium on Circuits and Systems (ISCAS), 2010, pp. 257-260 . https ://doi.org/10.1109/ISCAS.2010.5537908

Alaeddine Hmidi, Doctor in electronics and microelectronics.

He is currently a teacher of technical education in the secondary schools and institutes of the Tunisian Ministry of Education. Former assistant at the Institut Supérieur des Sciences Appliquées et de Technologie de Sousse - Tunisia and member of the Electronics and Microelectronics Laboratory of the FSM. His research work focuses on hardware architectures; FPGA, Deep Learning, Architectures and accelerators for image classification applications.

Jihene Malek, PhD in electronics and microelectronics.

Currently a lecturer at the Institut Supérieur des Sciences Appliquées et de Technologie de Sousse - Tunisia and a member of the Electronics and Microelectronics Laboratory at FSM. His current research interests include classification, combination of classifiers, feature extraction, high dimensional classification problems, fuzzy neural network sets, biomedical engineering, medical image processing and visualization, 3D reconstruction and hardware implementation.

yes I want morebooks!

Buy your books fast and straightforward online - at one of world's fastest growing online book stores! Environmentally sound due to Print-on-Demand technologies.

Buy your books online at
www.morebooks.shop

Kaufen Sie Ihre Bücher schnell und unkompliziert online – auf einer der am schnellsten wachsenden Buchhandelsplattformen weltweit! Dank Print-On-Demand umwelt- und ressourcenschonend produzi ert.

Bücher schneller online kaufen
www.morebooks.shop

KS OmniScriptum Publishing
Brivibas gatve 197
LV-1039 Riga, Latvia
Telefax: +371 686 204 55

info@omniscriptum.com
www.omniscriptum.com

Printed by Books on Demand GmbH, Norderstedt / Germany